Structure and Function of the Human Body

Structure and Function

Ruth Lundeen Memmler, M.D.

Professor Emeritus, Life Sciences;
formerly Coordinator, Health, Life Sciences and Nursing,
East Los Angeles College, Los Angeles

Illustrated by Anthony Ravielli

 17 92 J. B. Lippincott Company

of the *Human Body*
Second Edition

Dena Lin Wood, *R.N., B.S., P.H.N.*
Assistant Head Nurse, Los Angeles County—
University of Southern California
Medical Center, Los Angeles

PHILADELPHIA   **Toronto**

PAPERBOUND: ISBN 0-397-54194-5
CLOTHBOUND: ISBN 0-397-54203-8

Library of Congress Catalog Card Number 76-47605

Printed in the United States of America
     7   9   8   6

Library of Congress Cataloging in Publication Data
Memmler, Ruth Lundeen.
    Structure and function of the human body.
    Bibliography: p.
    Includes index.
    1. Human physiology. 2. Anatomy, Human. I. Wood,
Dena Lin, joint author. II. Title. [DNLM: 1. Anatomy.
2. Physiology. QD4 M533s]
QP36.M54 1977   612   76-47605
PAPERBOUND: ISBN 0-397-54194-5
CLOTHBOUND: ISBN 0-397-54203-8

Preface

As in the case of the first edition this concise text seeks to provide the beginning student with a basic understanding of what the parts of the body are and how they work together. To further clarify the basic facts of anatomy and physiology, new illustrations prepared by the outstanding anatomical artist, Anthony Ravielli, are included. The student should find them beautiful to look at and interesting to study. The authors and publishers believe they will supplement and enrich the text and may often tell the story better than words can. The emphasis is on normal structure (anatomy) and function (physiology).

Although the elemental building blocks are presented first, so that the student may proceed from the basic to the complex in a cohesive fashion, the order of presentation of the various chapters may be changed to fit a particular curriculum and to suit specific student needs. We believe this text lends itself to rearrangements that will suit both the instructors and the students. It has flexibility.

The pronunciation key of the first edition has been retained, and should be a time-saving guide in the study of new words. Students will find learning medical terms more productive if they will practice saying them aloud as they study. Knowing how to say a new word helps the student understand the word and remember its meaning. Students in all health fields need to be well acquainted with the terminology of anatomy and physiology so that it becomes a part of their everyday vocabulary.

The glossary has been updated to reflect the new content.

Since the metric system is being widely adopted, metric measurements are stated in the text whenever the size of an organ or a part is noted. Included also is an explanation of the Celsius (formerly called the centigrade) temperature scale.

Finally, the chapter summaries and the questions and problems sections have been retained. These various features should add greatly to the teaching-learning process this text seeks to augment.

To enhance the student's ability to apply his newly acquired knowledge of anatomy and physiology, an accompanying workbook is available. The workbook is closely correlated with the textbook, and affords numerous opportunities for self-testing.

We thank our husbands and friends who have been supportive and encouraging. We thank the many readers of the first edition who offered constructive comments. Particularly do we thank Edmond Stout, Associate Professor, Department of Education, and Lecturer in Biology, St. Joseph's College, Philadelphia, who painstakingly reviewed the entire manuscript and whose useful recommendations have contributed greatly to the text. We express our indebtedness to the J. B. Lippincott Company, particularly Bernice Heller, Editor, Nursing and Allied Health Sciences, and David T. Miller, Managing Editor, Nursing Department.

Ruth L. Memmler
Dena L. Wood

Contents

ONE

The General Plan of the Human Body

What Are Living Things Made Of?

According to a nursery rhyme children are made of sugar and spice, or perhaps of puppy dogs' tails, depending on which sex we are discussing. More accurately, the "stuff" of which all living things are made is called **protoplasm** (pro'to-plazm). This word is made up of 2 Greek words: *proto*, meaning "original," and *plasm*, meaning "substance." Chemically, protoplasm is composed of quite ordinary elements, such as carbon, oxygen, hydrogen, sulfur, nitrogen and phosphorus. There is nothing extraordinary, either, in the appearance of protoplasm; it looks very much like the white of an egg. Yet, nobody has been able to explain why protoplasm has that characteristic which we call life.

If the building material of all living things, both plants and animals, is protoplasm, the building blocks made of this are called **cells** (see Fig. 1.1). Cells vary a great deal in size. Something as small as a worm may be composed of millions of cells, yet we all are familiar with at least one of the larger kinds of cells, of which an egg is a perfectly good example. In fact, if we keep the egg in mind, the construction of the cell will be quite easy to visualize. Let us work our way from the outside to the center.

First comes the outer covering, called the **cell membrane**. Next is the main substance of the cell, the **cytoplasm** (si'to-plazm), which might be likened to the white of the egg. The cytoplasm contains water, food particles, pigment and other specialized materials. In the center of the cell, comparable with the egg yolk, is a globule called the **nucleus** (nu'kle-us), containing the chromatin network. The nucleus controls some of the activities of the cell, including its reproduction. Within the nucleus is still another tiny globule of matter called the **nucleolus** (nu-kle'o-lus), the function of which is related to reproduction. The unique ability of a cell to reproduce itself will be discussed in Chapter 2.

The scientific study of cells began with the invention by Antony van Leeuwenhoek of the microscope some 350 years ago. In time his single lens microscope was replaced by the modern compound microscope which has two sets of lenses. This is the type in use in most laboratories. In recent years a great boon to microbiologists has been the development of the electron microscope, which

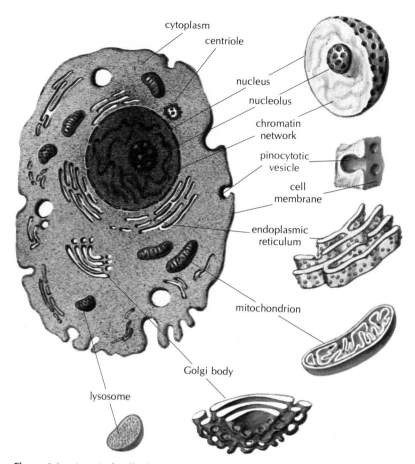

cytoplasm

centriole

nucleus

nucleolus

chromatin network

pinocytotic vesicle

cell membrane

endoplasmic reticulum

mitochondrion

Golgi body

lysosome

Figure 1.1. A typical cell. The nucleus is the control center. The organelles—the endoplasmic reticulum, the mitochondrion, the Golgi body, and the lysosomes—are the functional substances.

by a combination of magnification and enlargement of the resulting image affords magnification to one million times or more (Fig. 1.2).

The cell, then, is the basic unit of all life. When you study the causes of disease, you will encounter a number of primitive living things which are composed of but one cell. However, for the moment we shall confine our discussion to the human body, which is made up of many millions of cells. The body is composed of specialized groups of cells, the first of which are called **tissues**. Various tissues that together perform a single function form **organs**, and several organs and parts grouped together for certain functions form **systems**. The heart is an organ composed of muscle tissue, connective tissue and nerve tissue, all working together to pump blood. The heart and the blood vessels comprise the circulatory system.

Body Systems

The body systems have been variously stated to be nine, ten or eleven in number, depending on how much detail one wishes to include.

Here is one list of systems:

1. The **skeletal system**. The basic framework of the body is a system of over 200 bones with their joints, collectively known as the skeleton.

2. The **muscular system**. Body movements are due to the action of the muscles which are attached to the bones. Other types of muscles are

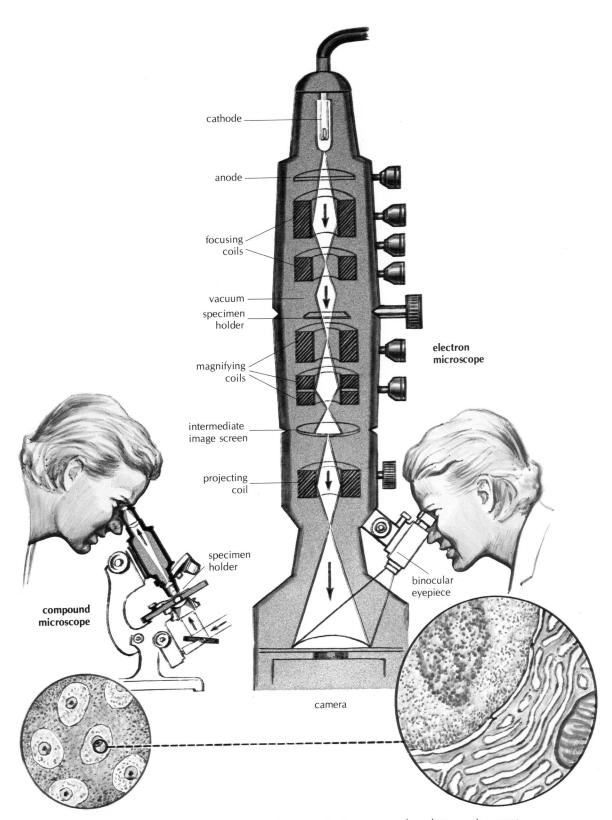

cathode

anode

focusing
coils

vacuum

specimen
holder

magnifying
coils

intermediate
image screen

projecting
coil

**electron
microscope**

specimen
holder

**compound
microscope**

binocular
eyepiece

camera

Figure 1.2. A simplified comparison of an optical microscope and an electron microscope.

present in the walls of such organs as the intestine and the heart.

3. The **circulatory system**. The heart, blood vessels, lymph vessels and lymph nodes all make up the system whereby blood is pumped to all the body tissues, bringing with it food, oxygen and other substances, and carrying away waste materials.

4. The **digestive system**. This system comprises all organs which have to do with taking in food and converting the useful parts of it into substances that the body cells can use. Examples of these organs are the mouth, the teeth, and the alimentary tract (esophagus, stomach, intestine and accessory organs such as the liver and the pancreas).

5. The **respiratory system**. This includes the lungs and the passages leading to and from them. The purpose of this system is to take in air, and from it extract oxygen which is then dissolved into the blood and conveyed to all the tissues. A waste product of the cells, carbon dioxide, is taken by the blood to the lungs, whence it is expelled to the outside air.

6. The **integumentary system**. The word "integument" (in-teg'u-ment) means "skin." The skin is considered by some authorities to be a separate body system. It includes the hair, nails, sweat and oil glands, and other related structures.

7. The **urinary system**. This is also called the excretory system. Its main components are the kidneys, the ureters, the bladder and the urethra. Its purpose is to filter out and rid the body of certain waste products taken by the blood from the cells. (Note that other waste products are removed via the digestive and the respiratory systems.)

8. The **nervous system**. The brain, the spinal cord and the nerves all make up this very complex system by which all parts of the body are controlled and coordinated. The organs of special sense (such as the eyes, ears, taste buds, and organs of smell), sometimes classed as a separate **sen**sory system, together with the sense of touch, receive stimuli from the outside world, which are then converted into impulses that are transmitted to the brain. The brain determines to a great extent the body's responses to messages from without and within, and in it occur such higher functions as memory and reasoning.

9. The **endocrine system**. A few scattered organs known as endocrine glands produce special substances called hormones, which regulate such body functions as growth, food utilization within the cells, and reproduction. Examples of endocrine glands are the thyroid and the pituitary glands.

10. The **reproductive system**. This system includes the external sex organs and all related inner structures which are concerned with the production of new individuals.

Directions in the Body

Because it would be awkward and incorrect to speak of bandaging the "southwest part" of the chest, a number of terms have been devised to designate specific regions and directions in the body. Some of the more important of these are listed as follows (note that they refer to the body in the "anatomic position"—upright with palms facing forward):

1. **Superior** is a relative term meaning "above" or "in a higher position." Its opposite, **inferior**, means "below" or "lower." The heart, for example, is superior to the intestine.

2. **Ventral** and **anterior** mean the same thing in humans: "located near the belly surface or front of the body." Their corresponding opposites, **dorsal** and **posterior**, refer to locations nearer the back.

3. **Cranial** means "near the head"; **caudal**, "near the sacral region of the spinal column" (i.e., where the tail is located in lower animals).

4. **Medial** means "near an imaginary plane that passes through the midline

of the body, dividing it into left and right portions." **Lateral**, its opposite, means "farther away from the midline," toward the side.

5. **Proximal** means "nearest the origin of a structure"; **distal**, "farthest from that point." For example, the part of your thumb where it joins your hand is its proximal region. The tip of the thumb is its distal region.

For convenience in visualizing the spatial relationships of various body structures to each other, anatomists have divided the body by means of three imaginary planes. Think of a body plane as a huge cleaver (see Fig. 1.3).

1. The **midsagittal** (mid-saj'i-tal) **plane**. If the cleaver were to cut the body in two down the middle in a fore-and-aft direction, separating it into right and left portions, the sections you would see would be midsagittal.

2. The **frontal plane**. If, instead of the above operation, the cleaver were held in line with the ears and then were brought down the middle of the body, creating a front and a rear portion, you would see a front (anterior or ventral) section and a rear (posterior or dorsal) section.

3. The **transverse plane**. If the cleaver blade were swung horizontally, it would divide the body into an upper (superior) part and a lower (inferior) portion. There could be many such cross sections, each of which is on a transverse plane.

Body Cavities

The body contains a few large internal spaces or **cavities** within which various organs are located. There are two groups of cavities: **dorsal** and **ventral** (see Fig. 1.4).

DORSAL CAVITIES

There are two dorsal cavities: (1) the **cranial cavity,** containing the brain; and (2) the **spinal cavity,** enclosing the spinal cord. Both of these cavities join, hence they are a continuous space.

VENTRAL CAVITIES

The ventral cavities are much larger than the dorsal ones. There are two ventral cavities: (1) the **thoracic cavity,** containing mainly the heart, the lungs and the large blood vessels, and (2) the **abdominal cavity.** This latter space is subdivided into two portions, one containing the stomach, most of the intestine, the kidneys, the liver, the gallbladder, the pancreas and the spleen; and a lower one called the **pelvis,** or pelvic cavity, in which are located the urinary bladder, the rectum and the internal parts of the reproductive system.

Unlike the dorsal cavities, the ventral cavities are not continuous. They are separated by a muscular partition, the **diaphragm** (di'ah-fram), the function of which is discussed in Chapter 13.

Regions in the abdominal cavity

Because the abdominal cavity is so large, it has been found helpful to divide it into nine regions. These are shown in Fig. 1.5. The three central regions are the **epigastrium** (ep-i-gas'tre-um), located just below the breastbone; the **umbilical** (um-bil'i-kal) **region** about the umbilicus (um-bil'i-kus), commonly called the navel; and the **hypogastric** (hi-po-gas'trik) **region**, the lowest of all of the midline regions. At each side are the right and left **hypochondriac** (hi-po-kon'dre-ak) regions, just below the ribs; then the right and left **lumbar** regions; and finally, the right and left **iliac,** or **inguinal** (in'gwi-nal), regions. A much simpler division into four quadrants (right upper, left upper, right lower, left lower) is now less frequently used.

The Metric System

Now that we have set the stage for further study of the body, its structure and its processes, a look at the metric system would be in order since it is rapidly replacing the present system of measurement in the United States. The drug industry and the health care industry already have converted to the metric system, so anyone who plans a career in health should be acquainted with metrics.

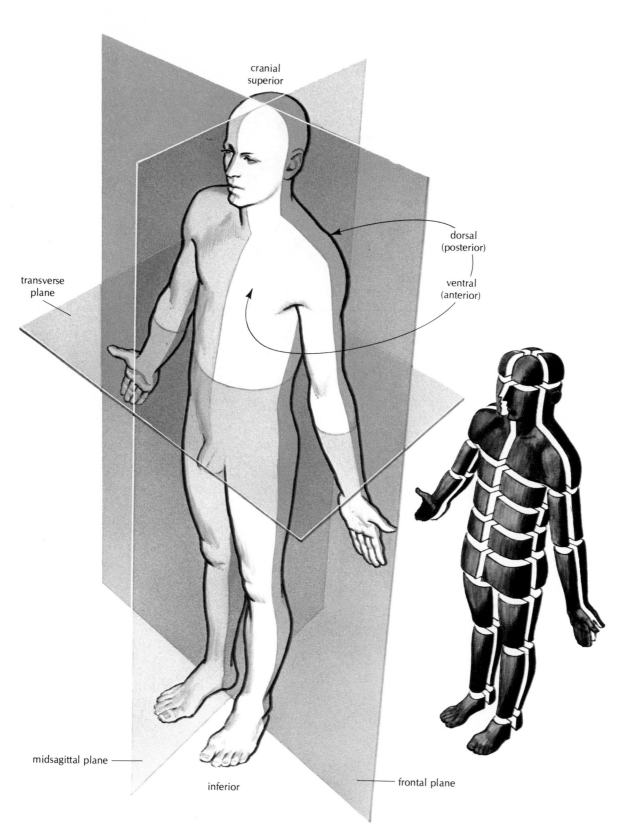

cranial
superior

dorsal
(posterior)

ventral
(anterior)

transverse
plane

midsagittal plane

inferior

frontal plane

Figure 1.3. Body planes and directions.

6

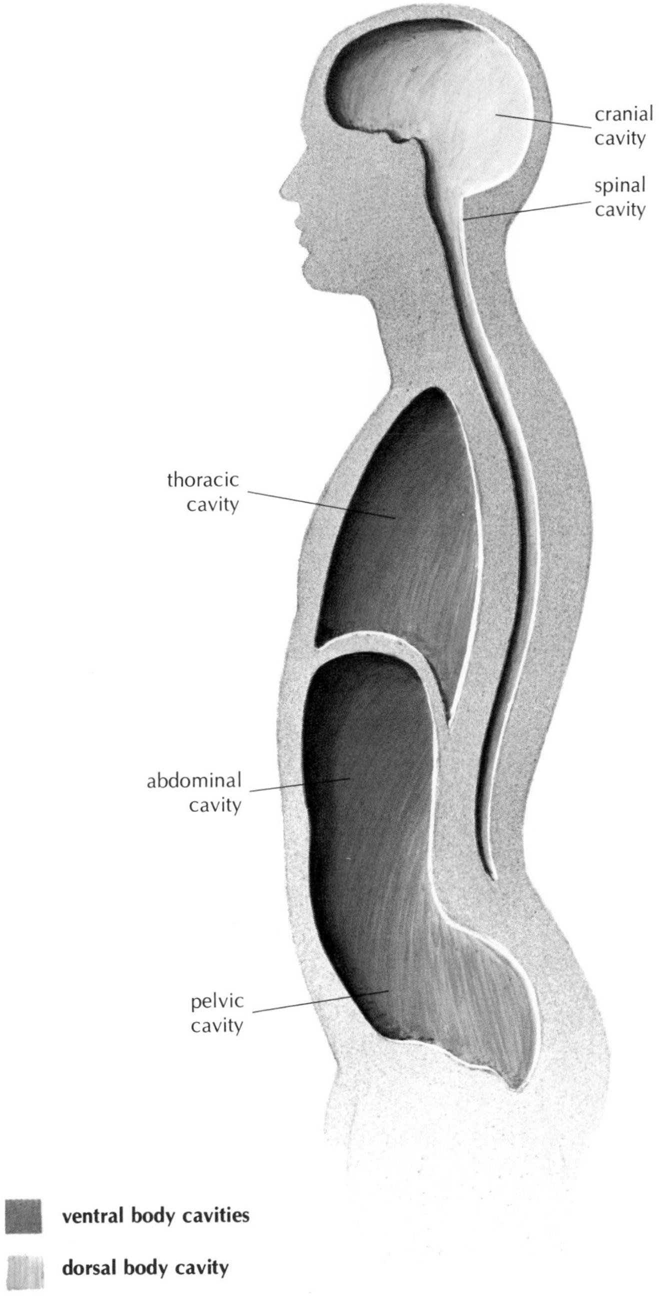
cranial
cavity
spinal
cavity
thoracic
cavity
abdominal
cavity
pelvic
cavity
ventral body cavities
dorsal body cavity

Figure 1.4. Side view of body cavities.

To use the metric system easily and correctly may require a bit of effort as is often the case with any new idea. Actually, you already know something about the metric system, because our monetary system is similar to it in that both are decimal systems. One hundred cents equals one dollar, 100 centimeters equals 1 meter. The first step is to learn the meanings of the following prefixes:

centi (1/100): 100 centimeters (cm.) = 1 meter (m.)

milli (1/1000): 1,000 millimeters (mm.) = 1 meter (m.)

kilo (1,000): 1,000 meters = a kilometer (km.) (kil′o-me-ter)

The metric system includes other mea-

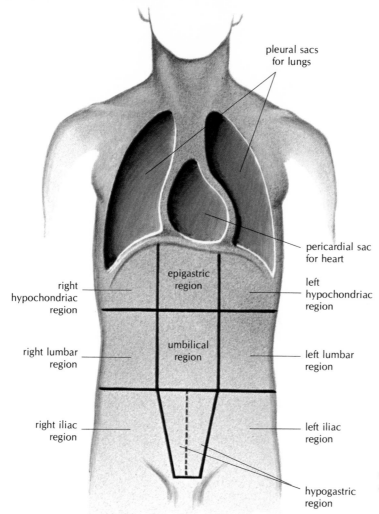

pleural sacs
for lungs

pericardial sac
for heart

right
hypochondriac
region

epigastric
region

left
hypochondriac
region

right lumbar
region

umbilical
region

left lumbar
region

right iliac
region

left iliac
region

hypogastric
region

Figure 1.5. Front view of body cavities and the regions of the abdomen.

surements, but these are the ones used most frequently. Millimeters, centimeters, meters and kilometers are linear measurements that correspond to inches, feet, yards and miles, respectively. In this text we have included the metric equivalents for inches, feet, etc., to indicate the size of an organ or part (Fig. 1.6). Some equivalents that may help you to appreciate the size of various body parts are:

1 mm. = 0.04 in., or 1 in. = 25 mm.
1 cm. = 0.4 in., or 1 in. = 2.5 cm.
1 m. = 3.3 ft., or 1 ft. = 30 cm.

The same prefixes are used as for the linear measurements. The meter is the standard for length, the gram for weights. Thirty grams is approximately equal to 1 ounce and 1 kilogram to 2.2 pounds. Drug dosages are usually stated in grams or milligrams. One thousand milligrams equals one gram; a 500 milligram (mg.) dosage would be the equivalent of 0.5 gram (g.) and 250 mg. is equal to 0.25 g.

The dosages of liquid medications are indicated as volume. The standard metric measurement for volume is the liter (le'ter). There are 1,000 milliliters (ml.) in a liter (l.). A liter is slightly greater than a quart, a liter being equal to 1.06 quarts. For

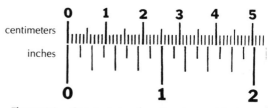

Figure 1.6. Comparison of centimeters and inches.

smaller quantities the milliliter (ml.) is used most of the time. There are 5 ml. in a teaspoon and 15 ml. in a tablespoon. A fluid ounce contains 30 ml.

The Celsius (centigrade) temperature scale now in use by most other countries as well as by scientists in this country is discussed in Chapter 4, Body Temperature and Its Regulation.

Summary

1. **Living matter.**
   A. Basic substance: protoplasm.
   B. Structural unit: cell.
   C. Principal parts of cell: cell membrane, cytoplasm, nucleus, nucleolus.
   D. Organization of body cells: tissues, organs, systems.
2. **Body systems:** skeletal, muscular, circulatory, digestive, respiratory, integumentary, urinary, nervous (and sensory), endocrine, reproductive.
3. **Body directions.**
   A. Superior, near head; inferior, away from head.
   B. Ventral (anterior), near belly; dorsal (posterior), near back.
   C. Cranial, near head; caudal, near end of spinal column.
   D. Medial, near midsagittal plane; lateral, toward side.
   E. Proximal, near origin; distal, distant from origin.

F. Body division by planes.
   (1) Midsagittal: left and right portions.
   (2) Frontal: front and rear portions.
   (3) Transverse: top and bottom portions.
4. **Body cavities.**
   A. Dorsal.
      (1) Cranial.
      (2) Spinal.
   B. Ventral.
      (1) Thoracic.
      (2) Abdominal.
         (a) 9 regions include epigastric, umbilical, hypogastric, right and left hypochondriac, right and left lumbar, and right and left iliac, or inguinal.
         (b) 4 quadrants (no longer extensively used).
   C. Dorsal cavities continuous, abdominal cavities separated by diaphragm.
5. **The metric system.**
   A logical decimal system.

Questions and Problems

1. Of what substance is living matter composed?
2. Define a cell. Name 4 main components of a typical cell.
3. Define: tissue, organ, body, system.
4. List the body systems, including a brief description of each with respect to its function.
5. List the opposite term for each of the following body functions: superior, ventral, anterior, cranial, medial, proximal. Define each item in the complete list.
6. What are the 3 main body planes? Explain the division of each.
7. Make a rough sketch of the 2 principal groups of body cavities, indicating the 9 divisions of the largest cavity.
8. Why should you learn the metric system? What are its advantages?

Cells, Tissues and Membranes

More About Cells

In Chapter 1 we learned that the cell is the fundamental building block of all life, no matter whether the living thing, plant or animal, is made of but one cell or many millions of them. The cell may live alone, or may be only one unit of a complex structure; but whatever its state, cell changes, including energy production and cell division (reproduction) go on constantly.

Careful microscopic studies of the cell protoplasm have revealed the presence of tiny structures called **organelles** (or-gah-nels') that are concerned with a variety of functions within the cell. Among these organelles are rod-shaped bodies called **mitochondria** (mit-o-kon'dre-ah). The mitochondria are responsible for the chemical combinations that result in the release of energy. This chemical activity involves the use of oxygen and nutrient materials and is accomplished by means of enzymes within the mitochondria.

Enzymes are complex proteins which act as catalytic agents, that is, they increase the speed of chemical reactions without being changed themselves. The vast majority of chemical reactions that go on in living things are catalyzed by enzymes. In addition to performing many functions within the cell itself, some enzymes are active outside of the cell in secretions as, for example, in the digestive juices. Enzymes act in a very specific manner, as if they are too choosey to have anything to do with substances other than the particular ones to which they are attracted. Among the most interesting of the substances found in cell protoplasm is DNA, or **deoxyribonucleic** (de-ok-se-ri-bo-nu-kle'ik acid (Fig. 2.1). It is found mostly in the cell nucleus. It is the chief component of the chromosomes, which contain the genes, the hereditary factors in each cell. The nucleus is the control center of the cell and DNA is a sort of master blueprint. The instructions contained in the DNA are carried to all parts of the body by the RNA, **ribonucleic** (ri-bo-nu-kle'ik) acid, whose formation DNA directs.

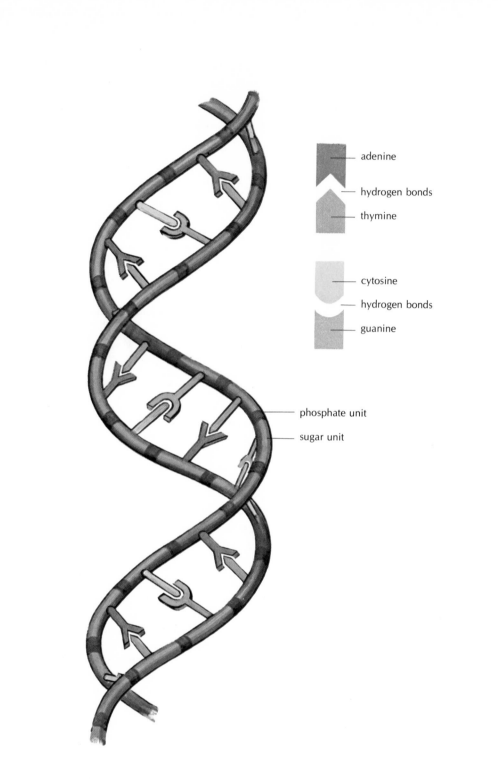

adenine

hydrogen bonds

thymine

cytosine

hydrogen bonds

guanine

phosphate unit

sugar unit

Figure 2.1. Schematic representation of the basic structure of a DNA molecule. Each structural unit consists of a phosphate group and a sugar group to which is attached a nitrogen base (adenine or thymine, cytosine or guanine). Adenine, thymine, cytosine and guanine are enzymes that "spell out" all the genetic instructions that control all activities of the cell.

When a cell has reached its limit of growth, it reproduces by dividing in two. The process of cell division is easy to visualize. First there is a division of the **centrosome** (sen'tro-som), a specialized structure just outside the nucleus of the cell containing two bodies called **centrioles** (sen'tri-ols). The divided centrioles then move to opposite sides of the cell, trailing threadlike substances which form a structure resembling a spindle stretched across the cell.

During every cell division the most spectacular changes take place in the nucleus. The chromatin material found in the usual resting cell changes to become rod-shaped bodies called **chromosomes** (kro'mo-soms). These contain the **genes** (jenes) which are responsible for the inherited traits of each cell and, eventually, those of the entire organism. The chromosomes split in half and separate, and each half replicates (repeats) itself. Half the chromosomes are drawn to each end of the cell, following the threadlike substances of the spindle that was formed by the centrioles.

Then the nucleus begins to elongate, becoming pinched in the middle until it resembles a dumbbell. The cell wall takes on the same shape. The midsection between the two halves of the dumbbell becomes smaller and smaller until finally the cell splits in two. We now have two cells where there was but one. Each of these two **daughter** cells (the original is called, logically enough,

the **parent** cell) usually receives exactly half the cell wall, half the cytoplasm, half the nucleus and other material of its parent. The name for this cell division is **mitosis** (mi-to'sis) (see Fig. 2.2). During the later phases of mitosis new pairs of centrioles are formed in preparation for the next cell division.

So far, so good; we have two new cells, but since each cell has only half the material that its parent contained, it is small and must grow before it is able to function or to reproduce itself in turn. Cells are alive, and all living things need food to grow and to produce energy; therefore the cell must receive nourishment somehow. This would seem to be a problem, since we learned that most cells are surrounded by a protective membrane. However, if the cell is bathed in a liquid containing dissolved food materials, an interesting thing happens: the liquid with the dissolved food particles passes through the cell membrane. Not only do the nutrient molecules pass in, but waste products pass out of the cell in the opposite direction. The membrane also keeps valuable protein and other substances from leaving the cell and prevents the admission of undesirable substances. For this reason, the cell membrane is classified as a **semipermeable** (sem-e-per'me-ah-b'l) membrane, being very selective in what it allows to enter and to leave the cell. It is permeable

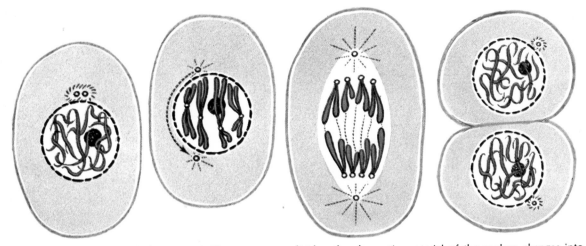

Figure 2.2. Mitosis (simplified sequence). The centrosome divides, the chromatin material of the nucleus changes into rod-shaped chromosomes, and two daughter cells form within the cell membrane.

or passable to some molecules but is impassable to others.

Water, having a tiny molecule, is always able to penetrate the membrane; whereas sugar (sucrose) cannot pass through because its molecule is too large—even though it is readily soluble. In the digestive process, sucrose is converted to glucose, which can diffuse through the membrane because its molecule is smaller.

A combination of various physical processes is responsible for the phenomenon of exchanges through the cell membrane or through the tissue membranes that are made up of a combination of many cells. Some of these are:

1. **Diffusion**, the constant movement of molecules from a region of relatively high concentration to one of lower concentration. Molecules, especially those in solution, tend to spread throughout an area until they tend to become equally concentrated in all parts of the container (see Fig. 2.3A and B).

2. **Osmosis**, the diffusion of water through a semipermeable membrane, with the molecules going from the less concentrated solution to the one that is more concentrated (the reverse of what happens in diffusion). The effect, however, still is a tendency to equalize the concentrations of the various substances in a given area (see Fig. 2.4).

Both the *size* of the molecules of the dissolved substance and the *nature* or *type* of membrane are determining factors in osmosis. Think of it this way: What would happen if you tried to throw beebees, marbles, ping-pong balls, tennis balls, handballs and basketballs through a tennis net? All of the beebees would get through, most of the marbles would, and some of the

Figure 2.3A. Diffusion of gaseous molecules throughout a given space. The bottle could contain perfume, a spray of some kind, a chlorine bleach, etc. In any case, there is a tendency for the molecules to spread throughout the area.

Figure 2.3B. Diffusion of a solid and a liquid. The solid diffuses into the water; the water diffuses into the solid; the molecules of solid tend to spread throughout the water.

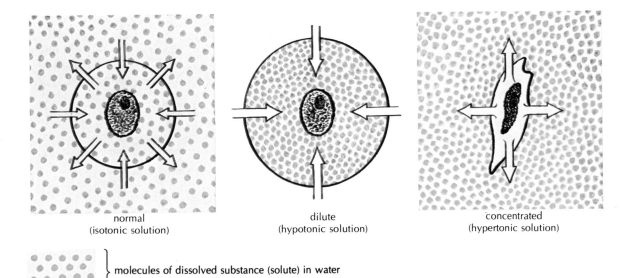

normal
(isotonic solution)

dilute
(hypotonic solution)

concentrated
(hypertonic solution)

molecules of dissolved substance (solute) in water

Figure 2.4. Osmosis: water molecules moving through a cell membrane into a solution of salts in high concentration. The normal saline solution has a concentration nearly the same as that inside the cell; the dilute solution causes the cell to swell and eventually to rupture because of the large number of water molecules moving into the cell; the concentrated solution causes the water molecules to move out of the cell, leaving it shrunken.

ping-pong balls would; but none of the tennis balls, handballs, or basketballs would (Fig. 2.5).

3. **Filtration**, the passage of water containing dissolved materials through a membrane as a result of a greater mechanical force on one side (see Fig. 2.6A and B). An example of filtration in the human body is the formation of urine in the microscopic func-

tional units of the kidney as described in Chapter 14.

Metabolism: Energy for the Cell

All the chemical reactions by which food is transformed for use by the cells is given the broad general name of metabolism. The metabolic process produces energy—the capacity for action.

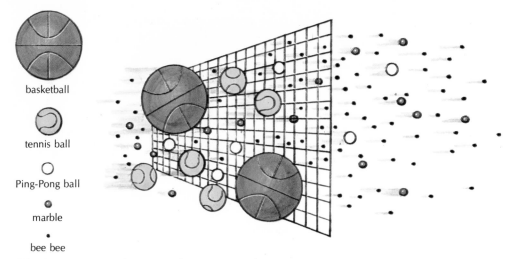

basketball

tennis ball

Ping-Pong ball

marble

bee bee

Figure 2.5. Representation of osmosis. The tennis net (the semi-permeable membrane) allows some of the balls (the dissolved materials) to pass through, but not others.

ENERGY FOODS

To be usable by cells, the ingested carbo-hydrates, fats, oils and proteins must be changed to simpler forms having smaller molecules, as described earlier. On the other hand, water, vitamins and minerals—the nonenergy group—are ready for immediate absorption and use without first being changed.

In the digestive process, all carbohy-drates are reduced to a simple sugar, glu-cose; fats and oils, to fatty acids and glyc-erol. Proteins are changed to amino acids. All of these end products are soluble in water, which serves as the transport me-dium to the cell's interior.

The glucose and fats (triglycerides) enter the mitochondria, where they oxidize (com-bine with oxygen) and release their energy. The amino acids are carried to the endo-plasmic reticulum, to be reassembled by the ribosomes to match the genetic code in the nucleus. Thus the synthesized protein matches exactly the protein already in the cell. This protein then is used to make new protein, in a never-ending cycle. The new protein causes the cell to grow to a certain size, which in normal health is regulated by a built-in mechanism. We can see why the mitochondria are often called the cellu-lar powerhouse.

What happens when the diet is deficient in needed protein? The endoplasmic reticu-lum becomes unable to assemble the needed protein for the cells; the undernourished cells cannot grow; the result is that the per-son is stunted in both physical and mental development. The effects of such deprivation are visible in some areas of the world where, because of a combination of igno-rance and poverty, the diet consists mainly of corn, rice and beans.

NONENERGY FOODS

There is a second group—not foodstuffs in the usual sense of the word—yet of criti-cal importance. Water provides no energy directly, yet is essential for life. By its fluid nature, it is the ideal medium for carrying nutrients to cells. Vitamins occur in minute amounts in most natural foods. They are needed for normal growth and regulation of metabolism. Minerals, which also occur in extremely small amounts in various foods, contribute to proper development of bones and teeth.

The Cellular Factory

A cell is very much like a tiny factory, complete with its own power plant. The cell consumes food, builds itself up, and generates heat and energy. As after any fire, waste products are left over. These are re-moved from the cell, to be carried away by the blood and other fluids. We have already investigated the cell's reproductive process.

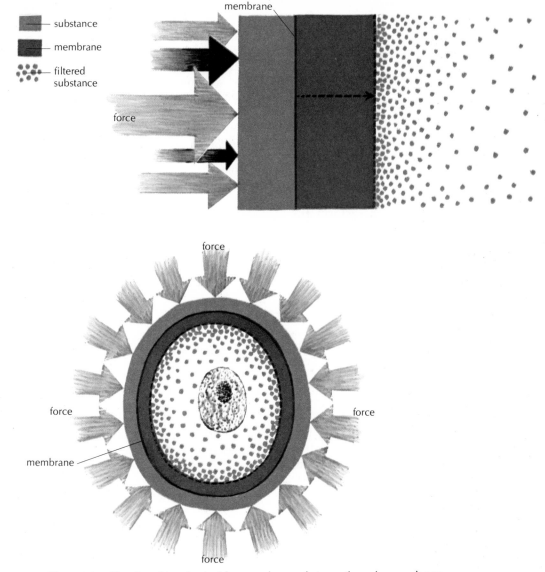

Figure 2.6   Filtration. A mechanical force pushes a substance through a membrane.

Two other general traits of cells remain. The first of these is the ability to adapt to changing outside conditions. The second is the ability to discharge fluids through the cell wall. This latter trait was, of course, partially covered in the discussion of waste disposal. However, some specialized cells also discharge other fluids called **secretions**. These are substances produced by chemical modification of materials taken into the cell. Often these secretions serve a useful purpose elsewhere in the body.

The foregoing material should be con-cluded with a few qualifications. Some kinds of cells reproduce more readily than others; certain types, if they die, are not replaced at all. On the other hand, many thousands of new cells are formed daily in the skin and other tissues to replace those destroyed by injury, disease or certain natural processes. As a person ages, characteristic changes in the overall activity of his body cells take place. One example of these is the slowing down of repair processes. The fracture of a bone, for example, takes considerably longer to heal in an aged person than in a young one.

At this point we are ready to proceed from the individual cell to the specialized groups of cells of which our bodies are made. Such groups are known as **tissues.**

On Tissues in General

Although the basic structure of cells, as well as certain behavior patterns, remains constant regardless of the type of cell under discussion, cells themselves vary enormously with respect to shape, size, color and specialty. Many cells, for example, are transparent; some of these form the "window" of the eye. Other cells may have extensions in the form of thin fibers over a yard long, as in the case of some nerve cells. Some produce secretions; others transmit electrical impulses.

Tissues are groups of cells similar in structure and substance and arranged in a characteristic pattern, and specialized for the performances of a specific task. In some ways the tissues in our bodies might be compared with the different materials which we use to clothe ourselves. Think for a moment of the great variety of materials employed in covering the body according to the degree of protection needed, the time of year, and so forth—wool, cotton, silk, rayon, leather and even straw. All of these have different properties; so do tissues.

Before we begin a more detailed discussion of tissues, it might be interesting to note the surprising fact that we are virtually living in water. The tissues are full of it. The cells of which tissues are made contain from 60 percent to 99 percent water. Gases, liquids and solids dissolve in this water. Chemical reactions that are necessary for proper body functioning are carried on much more readily in a watery solution. Substances which do not go into solution may be suspended in the various liquids of the body, many of which circulate; and thus they may be moved from place to place. Water is indispensable for cell life, and lack of water causes death more rapidly than the lack of any other dietary constituent.

The solution of water and other materials in which the tissues are bathed is slightly salty, an interesting reminder of the first living cells which originated in the sea.

This substance is called **tissue fluid.** Later, when we study the functions of the blood, we shall find out how food and oxygen manage to reach the cells of the tissues by way of the blood and the tissue fluid.

In connection with tissue fluid, it might be appropriate to mention here that an insufficiency of tissue fluid is called **dehydration,** and an abnormal accumulation of this fluid causes a condition called **edema** (e-de'mah), which shows up as a puffiness of the affected tissue.

Tissue Classification

The four main groups of tissue are:
1. **Epithelium**—forms glands, covers surfaces and lines cavities.
2. **Connective tissue**—holds all parts of the body in place.
3. **Nerve tissue**—conducts nerve impulses.
4. **Muscle tissue**—designed for power-producing contractions.

Blood sometimes is considered a sort of tissue, since it contains cells and performs many of the functions of tissues. However, the blood has so many other unique characteristics and purposes that an entire chapter will be devoted to it.

Epithelium

Epithelium (ep-e-the'le-um) forms a protective covering for the body and all its organs; in fact, it is the main tissue of the outer layer of the skin. It forms the lining of the intestinal tract, the respiratory and urinary passages, the blood vessels, the uterus and other body cavities.

Epithelium has many forms and many purposes, and the cells of which it is composed vary accordingly (Fig. 2.7). For instance, the cells of some kinds of epithelium produce secretions, such as **mucus** (mu'kus) (a clear, sticky fluid), digestive juices, perspiration and other substances. The digestive tract is lined with a special kind of epithelium whose cells not only produce secretions but are also designed to absorb digested foods. The air that we breathe passes over yet another form of epithelium that

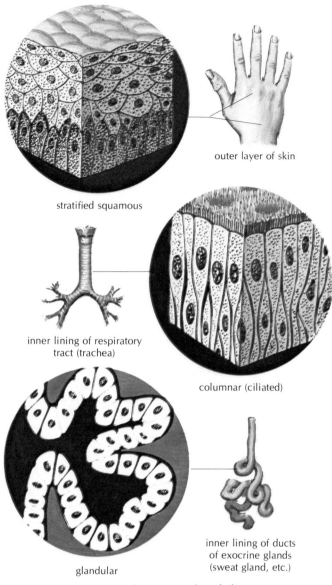

outer layer of skin

stratified squamous

inner lining of respiratory tract (trachea)

columnar (ciliated)

inner lining of ducts of exocrine glands (sweat gland, etc.)

glandular

Figure 2.7. Three types of epithelium.

lines the respiratory tract. This lining secretes mucus and is also provided with tiny hairlike projections called **cilia** (sil′e-ah). Together, the mucus and the cilia help to trap bits of dust and other foreign particles which could otherwise reach the lungs and damage them. Some organs, such as the urinary bladder, must vary a great deal in size during the course of their work; and for this purpose there is a special wrinkled, crepelike type of epithelium which is capable of great expansion, yet will return to its orig-inal form once the tension is relaxed—as when, in this case, the bladder is emptied. Certain areas of the epithelium that forms the outer layer of the skin are capable of modifying themselves for greater strength whenever they are subjected to unusual wear and tear; the growth of calluses is a good example of this.

Epithelium will repair itself very quickly if it is injured. If, for example, there is a cut, the cells near and around the wound immediately form daughter cells which grow until the cut is closed. Sometimes, however, particularly after repeated injury, abnormal growths will occur; and these are given the general name of **tumors**. Some tumors remain localized. Others spread; these are called **cancers**.

Connective Tissue

The supporting fabric of the organs and other parts of the body is connective tissue (Fig. 2.8). If we were able to dissolve all the tissues except connective tissue, we would still be able to recognize the contour of the parts and the organs of the entire body.

There are two distinct kinds of connective tissue, which are classified quite simply:
1. **Soft** connective tissues.
2. **Hard** connective tissues.

SOFT CONNECTIVE TISSUES

This group of connective tissues serves a number of different purposes. One group, called **adipose** (ad'e-pose) tissue, stores up fat for use by the body as a reserve food, a heat insulator and as padding for various structures. Another kind of soft connective tissue serves as a binding between organs, as well as a framework for some organs which are otherwise made of epithelium. There is yet another form of this tissue

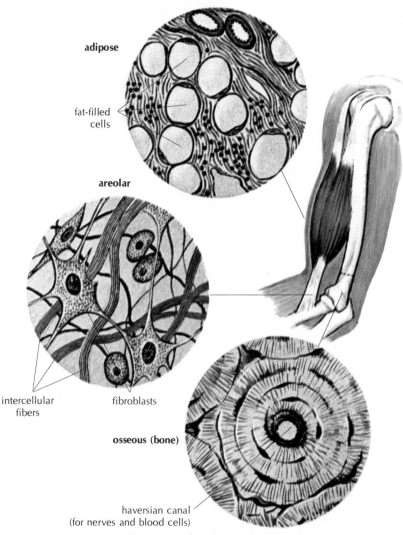

adipose

fat-filled cells

areolar

intercellular fibers

fibroblasts

osseous (bone)

haversian canal
(for nerves and blood cells)

Figure 2.8. Connective tissue.

which is particularly strong, being built up of fibers much like the strands of a cable. And, like a cable, this form of tissue serves to support certain organs which are subjected to powerful strains. A good example of this kind of tissue is a tendon. Yet another kind of soft connective tissue, called **neuroglia** (nu-rog'le-ah), is found in the brain and the spinal cord.

There is another interesting function of soft connective tissue, and this is its use by nature to repair muscle and nerve tissue as well as to repair connective tissue itself. A large gaping wound will require a correspondingly large growth of this new connective tissue, as will an infected wound; the new growth is called scar tissue. The process of repair includes stages in which new blood vessels are formed in the wound, followed by the growth of the scar tissue. An excessive development of the blood vessels in the early stages of repair may lead to the formation of so-called proud flesh. Normally, however, the blood vessels are gradually replaced by white fibrous connective tissue which forms the scar. Suturing (sewing) the edges of a clean wound together, as is done in the case of operative wounds, decreases the amount of scar tissue needed and hence reduces the size of the resulting scar. Such scar tissue may be stronger than the original tissue.

HARD CONNECTIVE TISSUES

The hard connective tissues, which, as the name suggests, are more solid than the other group, include cartilage and bone. Cartilage, popularly called gristle, is a tough, elastic and translucent material which is found in such places as between the segments of the spine and at the ends of the long bones. In these positions cartilage acts as a shock absorber as well as a bearing surface which reduces the friction between moving parts. Cartilage is found in other structures also, such as the nose, the ear, the epiglottis (the leaf-shaped structure below the throat) and other parts of the larynx, or "voice box."

The tissue of which bones are made, called **osseous** (os'e-us) tissue (see Fig. 2.8), is very much like cartilage in its cellular structure. In fact, the bones of the unborn baby, in the early stages of development, are (except for some of the skull bones) nothing but cartilage. However, gradually this tissue becomes impregnated with calcium salts; and since calcium is another word for lime, we see that a mineral deposit is going on which finally leaves the bones in their characteristically hard and stony state. Within the bones are nerves, blood vessels, bone-forming cells and a special form of tissue in which certain ingredients of the blood are manufactured.

A final comment on connective tissue in general is that, like epithelium, it repairs itself easily. In connective tissue, too, there may be abnormal growths of cells which form tumors.

Nerve Tissue

The human body is made up of countless structures both large and small, each of which (with a few notable exceptions) contributes something to the action of the whole organism. This aggregation of structures might be considered as an army, all of whose members must work together. In order that they may, there must be a central coordinating and order-giving agency somewhere; otherwise chaos would ensue. In the body this central agency is the brain. Each structure of the body is in direct communication with the brain by means of its own set of telephone wires, called nerves. The nerves from even the most remote parts of the body all come together and form a great trunk cable called the spinal cord, which in turn leads directly into the central switchboard of the brain. Here, messages come in and orders go out 24 hours a day. This entire communication system, brain and all, is made of nerve tissue.

The basic structural unit of nerve tissue is called a **neuron** (nu'ron) (Fig. 2.9). A neuron consists of a nerve cell body plus small branches like those of a tree. These are called **fibers**. One group of these fibers carries nerve impulses (i.e., messages) to the nerve cell body. Another group of fibers carries impulses away from the nerve cell body. Neurons can be tremendously long; for example, the one that reaches from the big toe to the brain spans many feet—and is composed of only one cell.

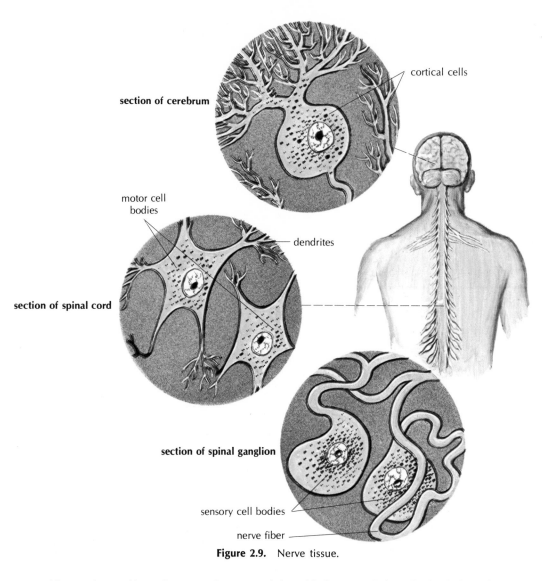

Figure 2.9. Nerve tissue.

Nerve tissue (i.e., clusters of neurons) is supported by ordinary connective tissue everywhere except in the brain. Here, the supporting tissue is of a special kind, and it is suspected of having a unique purpose of its own. So far, however, nobody knows exactly what this special function is.

All the nerves outside the brain and the spinal cord, called the **peripheral** (peh-rif'er-al) nerves, have a thin coating known as **neurilemma** (nu-re-lem'mah). Neurilemma is a part of the mechanism by which the peripheral nerves repair themselves when damaged. The brain and the spinal cord, on the other hand, have no neurilemma; so that

if they are injured, the injury is permanent. However, even in the peripheral nerves, repair is a slow and uncertain process.

Telephone wires are insulated to keep them from short-circuiting, and so are nerve fibers, which actually do transmit something very much like an electric current. The insulating material of nerve fibers is called **myelin** (mi'el-in), and groups of these fibers form "white matter," so-called because of the color of the covering; it is very much like fat in appearance and consistency. Not all nerves have myelin, however; some of the nerves of the system which controls the action of the glands, the smooth muscles (to be discussed shortly) and the heart do not have myelin. The cell bodies of all nerve cells also are uncovered (i.e., without myelin). Since all nerve cells are gray to begin with, and large collections of cell bodies are found in the brain, the great mass of brain tissue is popularly termed "gray matter."

Muscle Tissue

Muscle tissue, whatever its kind, is designed to produce power by a forcible contraction. The cells of muscle tissue are long and threadlike, and so are called muscle fibers. If a piece of well-cooked meat is pulled apart, small groups of these muscle fibers can be observed. Muscle tissue is usually classified as follows (Fig. 2.10):

1. **Skeletal muscle**, which combines with connective tissue to form the body muscles proper (to be discussed later along with the skeleton). These provide for the movement of the body. This type of tissue is also known as **voluntary** muscle, since it can be made to contract by those nerve impulses from the brain which originate from an act of will. In other words, in theory at least, any of your skeletal muscles can be made to contract as you want them to.

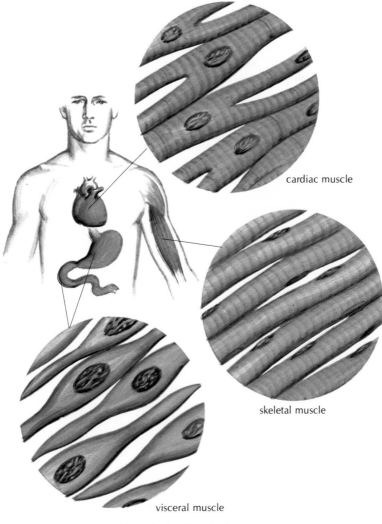

cardiac muscle

skeletal muscle

visceral muscle

Figure 2.10. Muscle tissue.

The next two groups of muscle tissue are known as **involuntary** muscle, since they typically contract independently of the will. In fact, most of the time we do not think of their actions at all. These are:

2. **Cardiac muscle**, which forms the bulk of the heart wall and is known also as **myocardium** (mi-o-kar'de-um). This is the muscle which produces the regular contractions known as heartbeats.

3. **Visceral muscle**, known also as **smooth** muscle, which forms the walls of the **viscera** (vis'er-ah), meaning the organs of the ventral body cavities (with the exception of the heart). Some examples of visceral muscles are those which move the food and waste materials along the digestive tract. Visceral muscles are found in other kinds of structures as well. Many tubular structures contain them, such as the blood vessels and the tubes which carry urine from the kidneys. Even certain structures at the bases of body hair have this type of muscle. When they contract, there results the skin condition that we call gooseflesh. Other types of visceral muscles will be taken up when we study the body systems.

Muscle tissue, like nerve tissue, repairs itself only with difficulty or not at all, once an injury has been sustained. These tissues when injured become replaced frequently with scar (connective) tissue.

Normal Growth

In the study of cells and tissues we gain some insight into the laws of growth. For reasons that still remain obscure, cells develop various forms, and those which are of the same kind congregate to form one of the basic tissues. These tissues become, in turn, the specialized organs. In the early stage of the body's development the cells multiply rapidly, and hence the body with all its structures grows in size until a point of maximum growth has been reached. From here on, cell division is not so rapid; it continues, however, at a rate sufficient to replace cells

which for one reason or other have become used up and discarded. This complete picture of growth is neat and logical: the tissues are maintained, and every cell and formation of cells has its purpose.

Characteristics of Membranes

Now that we have discussed the fundamental cell groupings—the tissues—we are ready to proceed to the next step and see in what ways the tissues are combined to form the actual body structures. The simplest of these tissue combinations are called **membranes**.

The word "membrane" means "any thin sheet of material which may separate two groups of substances." In our discussion of cells we encountered the word "membrane" for the first time, and noted that a membrane composed the wall of a cell through which various materials in solution (oxygen, food materials) could enter, and other substances (waste materials, secretions) could pass out. The cell wall is called the **cell membrane**.

In this chapter, however, we shall consider only those membranes which are made up of a multitude of cells—that is, of tissues. These are known as **tissue membranes**; but, for the sake of convenience, they shall henceforth be referred to simply as membranes.

Membranes are thin, skinlike layers of tissue. Their properties vary; some are fragile, and others are tough. Some are transparent while others are opaque, that is, they cannot be seen through. Membranes may serve as dividing partitions, or may line hollow organs and body cavities. They may contain secreting cells that produce lubricants which ease the movement of organs such as the heart and the movement of the joints. Other membranes serve to anchor various organs.

Kinds of Membranes

There are two broad categories of membranes. The first of these are the **epithelial membranes**, so-called because their outer surfaces are faced with epithelium. Their deep surfaces, however, have a layer of connective tissue, which strengthens the membrane. Epithelial membranes are in turn divided into two subgroups:

1. **Mucous** (mu'kus) **membranes,** which line tubes and other spaces that open to the outside of the body.
2. **Serous** (se'rus) **membranes,** which line closed cavities within the body.

The second category of membranes are known as fibrous **connective tissue membranes.** Unlike epithelial membranes, those of this group are composed entirely of connective tissue. This category of membranes also can be divided into two subgroups:

1. **Fascial** (fash'e-al) **membranes,** which serve to anchor and support the organs.
2. **Skeletal membranes,** which cover bone and cartilage.

Epithelial Membranes

Epithelial membranes are made of closely crowded active cells which manufacture lubricants and protect the deeper tissues from invasion by microorganisms. Mucous membranes produce a rather thick and sticky substance called mucus, while serous membranes secrete a much thinner lubricant. (Note that the adjective in each case contains an "o," while the nouns naming the secretion do not.)

In referring to the mucous membrane of a particular organ the noun **mucosa** (mu-ko'sah) may be used, while the special serous membrane covering an organ is called the **serosa** (se-ro'sah).

Mucous Membranes

Mucous membranes form extensive continuous linings in the digestive, the respiratory, the urinary, and the reproductive systems, all of which are connected with the outside of the body. They vary somewhat both in structure and function. The cells that line the nasal cavities and most parts of the respiratory tract are supplied with tiny hairlike extensions of the protoplasm, called cilia, which have been mentioned previously. The microscopic cilia move in a wavelike manner that forces the secretions outward away from the deeper parts of the lungs. In this way millions of pathogens, trapped in the sticky mucus, are prevented from causing harm. Ciliated epithelium is also found in certain tubes of both the male and the female reproductive systems.

The mucous membranes which line the digestive tract have their own special functions. For example, the mucous membrane of the stomach serves to protect the deeper tissues from the action of certain powerful digestive juices. If for some reason a portion of this membrane were injured, these juices would begin to digest a part of the stomach itself—which, incidentally, is exactly what happens in the case of peptic ulcers. Mucous membranes located farther along in this system are designed to absorb food materials which are then transported to all the cells of the body. But we are getting too far ahead in our story; other mucous membranes will be discussed as we encounter them.

Serous Membranes

Serous membranes, unlike mucous membranes, do not usually communicate with the outside of the body. This group lines the closed spaces known as body cavities. There are three main body cavities, and hence three serous membranes, which are:

1. The two **pleurae** (ploor'e) or pleuras (ploor'ahs), which form two separate sacs, one for each lung.
2. The **pericardium** (per-e-kar'de-um), which is a sac that covers the heart. It fits into a space in the chest between the two lungs.
3. The **peritoneum** (per-i-to-ne'um), which is much the largest, and which lines the abdominal cavity (see Fig. 2.11).

The epithelium covering serous membranes is of a special kind called **meso-thelium** (mes-o-the'le-um), which is smooth and glistening, and is lubricated so that movements of the organs can take place with a minimum of friction.

Serous membranes are so arranged that one layer forms the lining of the closed sac, while the other layer of the membrane covers the surface of the organs. Since the word **parietal** (pah-ri'e-tal) refers to a wall, the

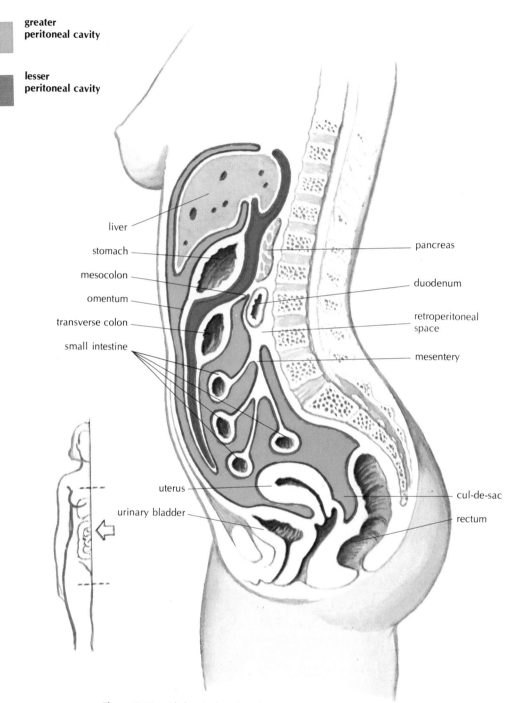

greater
peritoneal cavity

lesser
peritoneal cavity

liver

stomach

mesocolon

omentum

transverse colon

small intestine

uterus

urinary bladder

pancreas

duodenum

retroperitoneal
space

mesentery

cul-de-sac

rectum

Figure 2.11. Abdominal cavity showing peritoneum.

serous membrane attached to the wall of a cavity or sac is known as the parietal layer. There is, for example, parietal pleura lining the chest wall, and parietal pericardium lining the sac that encloses the heart. Because organs are called **viscera**, the membrane attached to the organs is the **visceral** layer. On the surface of the heart is visceral pericardium, while each lung surface is made up of visceral pleura.

Connective Tissue Membranes

Compared with epithelial membranes, connective tissue membranes are static, serving chiefly as retaining and supporting structures. These membranes, as has been mentioned, are divided into two subgroups, fascial and skeletal membranes.

FASCIAL MEMBRANES

The word "fascia" means "band"; hence fascial membranes are bands or sheets the purpose of which is to support the organs and hold them in place. An example of a fascial membrane is the continuous sheet of tissue which underlies the skin. This contains fat (adipose tissue or "padding") and is called the **superficial fascia**. "Superficial" refers to a surface; so the superficial fascia is closer than any other kind to the surface of the body.

As we penetrate more deeply into the body, we find examples of the **deep fascia**, which contains no fat and has many different purposes. Fascial membranes enclose the glands and the viscera; these envelopes are called **capsules**. Deep fascia covers and protects the muscle tissue; and these coverings are known as **muscle sheaths**. The blood vessels and the nerves also are sheathed with fascia; the brain and the spinal cord are encased in a multilayered covering called the **meninges** (me-nin'jez). In addition, fascia serves to anchor muscle tissue to structures such as the bones.

SKELETAL MEMBRANES

Skeletal membranes cover bones and cartilage. The membrane which covers the bones is known as **periosteum** (per-e-os'te-um), and that which covers cartilage is called **perichondrium** (per-e-kon'dre-um).

The cavities of the joints are lined with a membrane which is sometimes given a special classification among connective tissue membranes. This type is called the **synovial** (si-no've-al) membranes, and their particular purpose is to secrete a lubricating fluid which reduces the friction between the ends of bones, thus permitting free movement of the joints.

With this we conclude our brief introduction to membranes. As we study each system in turn, other membranes will be encountered. They may have unfamiliar names, but they will be either epithelial or connective tissue membranes; and we shall also be familiar with their general locations. In short, they will be easy to recognize and remember.

Summary

1. **Characteristics of cells.**
   A. The building block of living things.
   B. Organelles are responsible for various functions within the cell.
   C. Enzymes are catalytic agents.
   D. DNA and RNA.
   E. The cell reproduces by mitosis.
   F. Physical and biological processes bring materials through the semipermeable cell wall.
   G. Chemical action within the cell is called metabolism.

2. **Tissues.**
   A. Tissues made of specialized cells.
   B. Tissue compares with cloth; properties vary with function.
   C. Tissue fluid is mostly water.
   D. Insufficiency of fluid in tissue: dehydration. Too much: edema.

3. **Tissue classification.**
   A. Epithelium.
   B. Connective tissue.
   C. Nerve tissue.
   D. Muscle tissue.

4. **Epithelium.**
   A. Forms protective covering of the body and its organs.
   B. Forms lining of the intestinal tract, respiratory and urinary passages, blood vessels, uterus and other body cavities.
   C. May produce secretions.
   D. May have cilia or other special characteristics.
   E. Some types are wrinkled.
   F. Repairs itself quickly and easily.

5. **Connective tissue.**
   A. Supports organs and other body structures.
   B. Divided into types:
      (1) Soft. Stores fat, binds organs, forms

organ framework, supports organs with heavy strains, forms scar tissue.
  (2) Hard. Bones and cartilage.
C. Repairs itself easily and can grow tumors.

6. Nerve tissue.
A. Basic structure is neuron.
B. Composes coordinating and communication systems of the body.
C. Neurons vary greatly in length.
D. Neurilemma helps nerves to repair themselves. Peripheral nerves have neurilemma, brain and spinal cord do not, so are incapable of repair.
E. Repair of nerves even with neurilemma is slow and uncertain.
F. Myelin insulates some nerves.

7. Muscle tissue.
A. Primary purpose is to provide forcible contractions.
B. Fiberlike cells.
C. Three kinds.
  (1) Skeletal. Forms the body muscles proper, also called voluntary muscle.
  (2) Cardiac. Contracts regularly to produce heartbeat. An involuntary muscle.
  (3) Visceral. Known also as smooth muscle. Forms the walls of internal organs (heart excepted) including tubular structures.
D. Repairs itself with difficulty or not at all. Injured tissue may be replaced with scar tissue.

8. Characteristics of membranes.
A. Simplest combinations of tissue.
B. Thin, skinlike layers of tissue.
C. Secrete substances, line cavities, support organs.

9. Kinds of membranes.
A. Epithelial: outer surface is epithelium; deep surface is connective tissue.
B. Connective tissue: composed entirely of connective tissue.

10. Epithelial membranes.
A. Mucous membranes: secrete mucus, line passages which communicate with the outside of the body.
B. Serous membranes.
  (1) Characteristics: are covered with mesothelium, line body cavities, are lubricated thinly, have parietal and visceral layers.
  (2) Three serous membranes: pleurae (line lung cavities); pericardium (heart sac); peritoneum (lines abdominal cavity).

11. Connective tissue membranes.
A. Characteristics: static; retaining and supporting structures.
B. Kinds.
  (1) Fascial membranes: superficial fascia has fat, is below dermis of skin; deep fascia forms capsules, muscle sheaths; sheaths for nerves and blood vessels; anchors muscle fibers to bones.
  (2) Skeletal membranes: cover bones and cartilage. Periosteum (bone covering); perichondrium (cartilage covering). Synovial membranes line joint cavities and secrete joint lubricant.

Questions and Problems

1. Define each of the following: organelle, mitochondria, enzyme, semipermeable membrane, DNA and RNA.
2. Outline the various stages of cell division. What is the name of this process?
3. What are some examples of processes that are responsible for the exchange of materials through membranes?
4. What goes on inside the cell once it receives these materials? Give the name for this process.
5. Define a tissue. Give a few general characteristics of tissues.
6. Define epithelium and give 2 examples. How easily does it repair itself?
7. Define connective tissue. Name the main kinds of connective tissue and give an example of each. How easily does it repair itself?
8. What is the main purpose of nerve tissue? What is its basic structural unit called?
9. Define neurilemma. Where is it present or absent?
10. Define myelin. Where is it found?
11. How easily does nervous tissue repair itself?
12. Name the 3 kinds of muscle tissue and give an example of each.
13. What is the difference between voluntary and involuntary muscle?
14. How easily does muscle tissue repair itself?

15. What does the word "membrane" mean, generally speaking?
16. What is the general name for the membranes with which this chapter deals?
17. Name some general characteristics of membranes.
18. What are the 2 broad categories of membranes?
19. What are some general characteristics of epithelial membranes?
20. Name the 2 subgroups of epithelial membranes.
21. Give some characteristics of mucous membranes and name 2 examples of them.
22. Name some characteristics of serous membranes.
23. Name the 3 serous membranes and locate each.
24. What is the name for the kind of epithelium that covers serous membranes?
25. Name the 2 layers of serous membranes and tell what each means.
26. Give some general characteristics of connective tissue membranes.
27. Name the 2 subgroups of connective tissue membranes.
28. What are the main purposes of fascial membranes?
29. Name 2 kinds of fascia.
30. Give 3 examples of deep fascia.
31. What are the main purposes of skeletal membranes?
32. Name 3 examples of skeletal membranes.

The Blood

Purposes of the Blood

It was noted in Chapter 2 that the blood is sometimes classified as a tissue, since nearly half of it is made up of cells. However, because blood has so many unique functions, it deserves to be studied by itself. But the blood should be thought of as being akin to the tissues in that it is one of the primary materials of the body; and without some acquaintance with the composition and the functions of the blood, we could not understand the workings of the body's organ systems.

Blood is a thick (viscous) fluid which varies in color from bright scarlet to a darker brownish red, depending on how much oxygen it is carrying. The average adult has about 6 quarts of blood in his body (approximately 5.7 liters).

The blood has two main purposes. These are:

1. **Transportation.** Oxygen from the air that is breathed into the lungs is absorbed by the blood through the thin lung tissue and carried to all the tissues of the body. Carbon dioxide, a waste product of cell metabolism, is carried by the blood from the tissues to the lungs, where it is breathed out. The blood carries food materials from the intestine to all the body tissues, and waste products are transported by the blood to the kidneys for excre-

tion. Special secretions called hormones, whose purpose it is to regulate the body's growth, development and normal functions, are transported by the blood from their organs of origin to their various destinations. The blood also transmits heat that is generated in the muscles to other parts of the body, thus aiding in the regulation of body temperature. The blood transports certain mineral salts which maintain what is called the acid-base balance of the body (this will be explained later), build up bones and perform other functions.

2. **To Combat Infections.** Certain materials in the blood are among the body's great defenses against pathogenic invasions; other blood constituents are concerned with maintaining the body's immunity to disease.

Blood Constituents

The blood is composed of two prime elements. These are:

1. The liquid element, called **plasma.**
2. The so-called **formed elements**, which are cells and products of cells (Fig. 3.1). The formed elements also are called **corpuscles** (kor'pus'ls) and are grouped as follows:
   A. **Erythrocytes** (e-rith'ro-sites)—red blood cells (*erythro*-red).

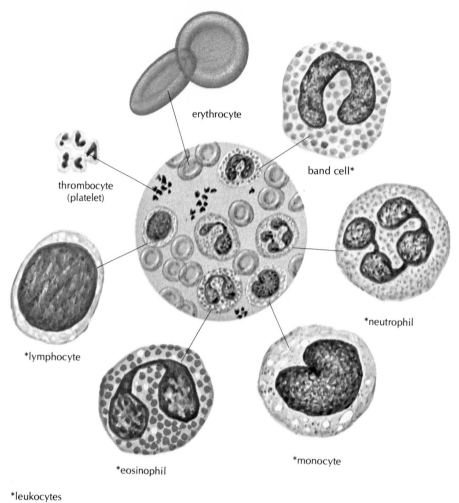

erythrocyte

band cell*

thrombocyte
(platelet)

*neutrophil

*lymphocyte

*monocyte

*eosinophil

*leukocytes

Figure 3.1 Blood cells.

B. **Leukocytes** (lu'ko-sites)—white blood cells (*leuko*-white).
C. **Platelets**—particles that bring about the process of clotting. These are probably not cells, but cell products. Platelets are called by another name also: **thrombocytes** (throm'bo-sites).

BLOOD PLASMA

Well over half of the total volume of blood is plasma; and plasma itself is approximately 90 percent water. The remaining part of plasma contains around 100 different substances dissolved or suspended in this water. The plasma content varies somewhat, since the blood carries substances to and from organs which use some of this material and add others. However, in the case of many substances there is a tendency to maintain a certain constant level. Glucose (simple sugar), for example, is kept at an average of about one-tenth of 1 percent solution (or about one part in a thousand). This is possible partly because glucose is stored in certain cells, especially those of the liver and muscles, and is later released as it is used by the tissues to generate energy.

After water, the next largest percentage of material of which the plasma is composed is called **protein**. Proteins are chemical compounds which are the principal constituents of protoplasm; they are essential to the growth and the rebuilding of body tissues—hence their importance. The proteins include such substances as:

1. The amino acids absorbed by the capillaries of the villi. These usually are described as protein building blocks.
2. Such vital compounds as antibodies that combat infection.
3. Certain proteins that figure in blood clotting.

Another substance in plasma belongs to a group with the collective name of **carbohydrates** (kor-bo-hi'drates). The principal form of carbohydrate found in the plasma is glucose, also absorbed by the capillaries of the villi; as we saw, it is stored up as reserve food or released to supply energy.

Lipids (lip'ids) also constitute a small percentage of the blood plasma. Lipids are substances that include fats, among other things. Some lipids also are stored (in the form of fat) or carried to the tissues to supply energy.

Another important ingredient of plasma is the group known as **mineral salts**. Mineral salts include calcium and sodium compounds, various carbonates and phosphates, and potassium and magnesium salts. All these salts are very important to the functioning of the cells. Some mineral salts are essential to the formation of bones (calcium and phosphorus). Other mineral salts are essential to the production of hormones in certain glands (e.g., iodine in the thyroid gland). Iron, another mineral of this group, is necessary for the transportation of gases (oxygen and carbon dioxide) by the red cells. Still other salts serve to maintain the body's acid-base balance.

Other materials in the blood plasma include waste products, hormones and gases.

THE FORMED ELEMENTS, OR CORPUSCLES

The word "corpuscle" means "little body." We shall investigate each of the three types of corpuscles in turn.

Erythrocytes

Erythrocytes, the red cells, are tiny disk-shaped bodies with a central area that is thinner than the edges. They are different from other cells in that the mature form found in the circulating blood does not have a nucleus. These cells live a much shorter time than most other cells of the body, some of which last a lifetime. One purpose of the red cells is to carry oxygen from the lungs to the tissues. This is accomplished through the main ingredient of the red cells, which is called **hemoglobin** (he-mo-glo'bin). It is the hemoglobin which absorbs this gas and gives the blood its characteristic red color. The more oxygen that is carried by the hemoglobin, the brighter the red color of the blood. Therefore, the blood that goes from the lungs, through the arteries, to the tissues is bright red because it carries a brand-new supply of oxygen. On the other hand, the blood that returns from the tissues, via the veins, and back to the lungs is a much darker red, since it has given up much of its oxygen. The red blood cells also carry the carbon dioxide from tissues back to the lungs for expiration.

The erythrocytes are by far the most numerous of the corpuscles, averaging from 4.5 to 5 million per cubic millimeter of blood.

Leukocytes

The leukocytes, or white blood cells, are very different from the erythrocytes in appearance, quantity and function. They contain nuclei of varying shapes and sizes, and the cell itself is shaped like a ball. Leukocytes are outnumbered by red cells by 700 to 1. Whereas the red cells have a definite color, the leukocytes tend to be colorless. The white cells have many different divisions, but for the moment it is sufficient for us to know that the most important function of leukocytes is to destroy certain pathogens. At any time that pathogens enter the tissues, as through a wound, the white blood cells are attracted to that area. They leave the blood vessels through their walls and proceed by what is called **ameboid** (ah-me'boid) or amebalike motion to the area of infection. There they engulf so many of the pathogens that very often they themselves die and disintegrate (see Fig. 3.2). A collection of dead and living bacteria, together with dead as well as living leukocytes, forms pus; and a collection of pus localized in one area is known as an abscess. Note: The combining form "leuko," meaning "white," may also be written "leuco."

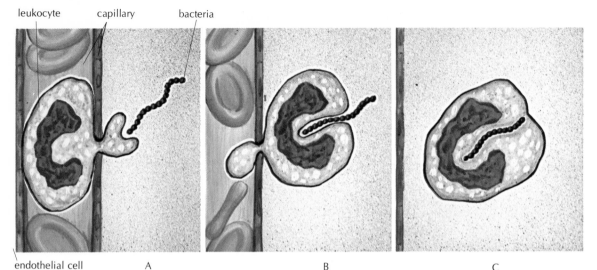

leukocyte   capillary   bacteria

endothelial cell   A   B   C

Figure 3.2. A, a white blood cell squeezes through a capillary wall in the region of an infection. B and C, the white cell engulfs the bacteria. This process, called phagocytosis, is a part of the body's mechanism for fighting infection.

Platelets

If it were not for the platelets, or thrombocytes, we would not last very long because the slightest cut would prove fatal; we would bleed to death. The platelets, then, are essential to blood clotting, or coagulation. They measure 250,000 to 300,000 per cubic millimeter. Platelets are not believed to be cells in themselves, but are probably fragments of cells. When blood is shed, or else comes in contact with any tissue other than that which normally carries blood, the platelets immediately disintegrate and release a chemical which reacts with a protein called **fibrinogen** (fi-brin'o-jen) which is manufactured in the liver and circulates in the plasma. The fibrinogen changes from a liquid to a solid mass called **fibrin**, which forms the clot. (The process by which soluble fibrinogen is changed to insoluble fibrin is complex, involving the enzyme thrombin, calcium ions and other substances).

Origin of the Corpuscles

The erythrocytes are formed in the red bone marrow, which is the connective tissue found inside the numerous small spaces of the spongy part of all bones. However, the red marrow in which the blood corpuscles are formed should not be confused with the yellow fatty kind which all of us have seen in soup bones, and which has its counterpart in the central cavities of the long bones of humans. Red marrow is found only in the ends of the long bones and in the mass of others.

As has been noted, the red cell as it normally appears in the blood has no nucleus. This was not always true; oddly enough, when red cells are being formed in the marrow, they have nuclei. However, the red cell must lose its nucleus before it is considered mature and ready to be released into the blood stream.

The leukocytes are for the most part born in the bone marrow, as the red cells are. However, we have noted that there are many different kinds of leukocytes, and some have another origin. One group of leukocytes, known as **lymphocytes** (lim'fo-sites), originate not in the marrow but in the lymph nodes and other lymphoid tissues. A discussion of lymph tissues will be reserved for a later chapter.

We have seen that when any pathogen enters the tissues, the leukocytes are attracted to the site. At the same time the leukocyte-forming tissue goes into emergency war production, so to speak, with the result that the number of leukocytes in the blood is enormously increased. Therefore, if in the course of a blood examination an abnormally large number of white cells are seen to be present, this sometimes is indicative of an infection somewhere. An abnormally small number of white cells is a characteristic sign of a different category of disease.

The platelets are believed to originate in the red marrow, as are nearly all the other corpuscles. They are fragments of certain giant cells, called **megakaryocytes** (meg-ah-kar'e-o-sites), which are formed in the marrow.

Blood Clotting

Blood clotting, or coagulation, is a protective device that prevents blood loss when a vessel is ruptured by injury. Among the many substances involved in the process are some whose function is to prevent clotting—anticoagulants—and some whose function is to promote coagulation—procoagulants. Under normal conditions the substances that prevent clotting prevail; with injury the procoagulants are activated and a clot forms.

Basically, the general scheme of coagulation is as follows:

1. The injured tissues release **thromboplastin**, (throm-bo-plas'tin) a substance that triggers the clotting mechanism.
2. Thromboplastin reacts with certain protein factors and calcium ions to form prothrombin activator, which in turn reacts with calcium ions to convert the prothrombin to **thrombin.**
3. Thrombin converts soluble fibrinogen into insoluble fibrin. Fibrin is a network of threads that entraps red blood cells and thereby forms a **clot.**

Blood Typing and Transfusions

BLOOD GROUPS

If for some reason the amount of blood in the body is severely reduced, through **hemorrhage** (hem'or-ij), which means copious bleeding, or through disease, the body cells suffer from lack of food and oxygen. The obvious measure to take in such an emergency is the injection of blood from another person into the veins of the patient, which is called a **transfusion.** However, before other aspects of blood transfusion are discussed, it should be pointed out that not just anybody will do as a blood donor.

The blood of some persons is not compatible with that of others. The plasma of one person may contain substances that will damage the red blood cells of another. The red cells of the donor's blood may become clumped or held together in bunches, a process called **agglutination** (ah-gloo-ti-na'shun), by a substance in the patient's blood. It may happen also that the cells of the donor are dissolved or go into solution; they then are said to be **hemolyzed** (he'mo-lizd), a most dangerous condition. These reactions are determined largely by the type of protein in the red cell. Four blood types have thus been recognized and are referred to as the A, B, AB and O types. The AB person has two proteins which may be agglutinated, while the O type has neither of these proteins. Because of the lack of these particular proteins, the O type person is known as a universal donor, that is, he may safely give blood to anyone. About 40 percent of people have type O blood (see Fig. 3.3). Whatever the type of blood a person may have, he can usually give blood safely to another with the same blood type. In all cases, before a transfusion is given, determination of the blood type and a further check for incompatibility always should be made.

THE RH FACTOR

About 85 percent of the population have another red cell protein called the **Rh factor.** Such individuals are said to be **Rh positive.** A minority of about 15 percent lack this protein and are said to be **Rh negative.** If Rh positive blood is given (say by transfusion) to an Rh negative person, he may become **sensitized** to the protein of Rh positive blood. That is, the blood of this person may produce counteracting substances called **antibodies** which in turn will destroy the erythrocytes contained in the "foreign" Rh positive blood (called **antigens**). A mother who is Rh negative may become sensitized

anti-B serum   anti-A serum

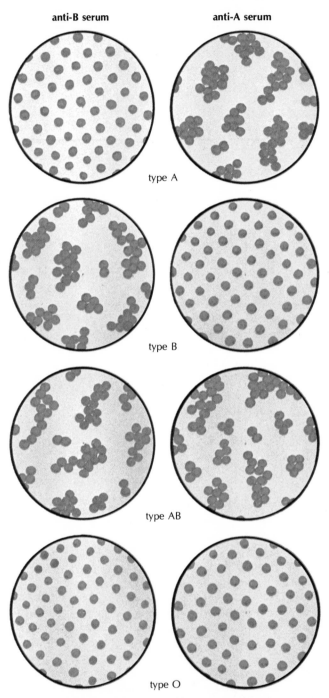

type A

type B

type AB

type O

Figure 3.3. Blood typing. Red cells in type A blood are agglutinated (clumped) by anti-A serum; those in type B blood are agglutinated by anti-B serum. Type AB blood cells are agglutinated by both serums; and type O blood is not agglutinated by either serum. Blood serum is the watery part of the blood that remains after the clot has been removed.

by proteins from an Rh positive baby (this factor having been inherited from the father), if such proteins find their way into the mother's circulation before or during childbirth. During a later pregnancy the mother's antibodies may pass from her blood into the blood of this second or third infant and there cause destruction of red cells. This results in a condition called **erythroblastosis fetalis** (e-rith-ro-blas-to'sis fe-ta'lis). The child may be born dead (stillborn). If the infant is alive, replacement transfusions are begun at once. Rh negative blood from a female donor seems to give the best results.

It is now possible to prevent fetal erythroblastosis by administering a globulin called Rho(D) immune globulin, human. This substance prevents the formation of active antibodies in an Rh negative mother who has received Rh positive blood via the placenta as a result of delivering an Rh positive infant. The globulin is also effective in cases of blood transfusion in which an Rh negative person has received Rh positive blood. Two names by which the substance is popularly known are Gamulin and RhoGam. This globulin should be given following a miscarriage in which the aborted fetus is likely to be Rh positive and the mother is Rh negative.

DETERMINATION OF PARENTHOOD

Blood grouping is inherited in a way that is somewhat like the inheritance of eye and hair color, following what is known as **Mendelian** (men-de'le-an) laws, after Mendel, who first formulated the laws of heredity. Therefore, in certain cases, where the identity of the father is sought, blood group studies of the child and the possible father are made. Unfortunately they provide only negative types of evidence, since such findings tend to prove who could not have been the father, and who might be; but they do not prove conclusively which one actually is the father.

BLOOD BANKS

Blood can be bottled and kept available for emergencies in blood banks. In order to keep the blood from clotting, **sodium citrate** (sit'rate or si'trate) in solution is added. This blood may then be stored for a number of days, usually not more than one to a maximum of three weeks. Such blood storage is especially important in times of disaster and during wartime. The supplies of blood in the bank are dated, and these dates are noted before the transfusion is given to avoid giving blood in which red cells may have disintegrated. Since, as we saw, about 40 percent of persons have type O blood, and since this type may be used for all types of patients (in an emergency), it is especially important to have larger supplies of this type in the blood bank. In any case, the patient's blood must be crossmatched with the blood to be used before the transfusion is begun.

Uses of Blood Derivatives

Blood is capable of being broken down into its various components and the substances derived from it used for a number of purposes. One of the more common of these processes is to separate the blood plasma from the formed elements. This is accomplished by means of a **centrifuge** (sen'tre-fuge), which is a machine that spins a quantity of blood around in a circle at high speed. If you imagine a weight tied to the end of a string, and think of spinning the weight around in a circle, you will understand how a centrifuge works. There is a force which tends to pull the weight outward. When the container of blood is spun rapidly, that same force will "pull" all the formed elements of the blood into a clump at the bottom of the container, separating them from the plasma, which can simply be poured off.

The blood plasma thus derived is a very useful substance. It may be given as an emergency measure to combat shock and to replace blood volume. The water can be removed, leaving the solids which can be stored in the dry state for a considerable length of time. Later, sterile distilled water may be added in order to reconstitute the plasma; it can then be given to treat an injured person, as, for example, in situations that do not make blood typing and the use of whole blood possible (on battlefields or in mass disasters). Since the red cells have been removed, there can be no incompatibility problems; plasma can be given to anyone.

Serum is another blood derivative. We all have observed that if a blood clot is removed (from a cut, for example) a watery fluid remains. This watery fluid is serum, and it is nothing more than plasma from which fibrinogen has been removed through the process of clotting. Serum may be derived from the blood of specially treated animals, and then injected into humans in order to produce an immunity to certain diseases.

Other blood fractions used for treatment purposes include the cells which are left after the plasma has been removed. These cells are made into a paste that is used to aid in the healing of ulcers, burns and other injuries.

Gamma globulin is a plasma protein that develops in certain tissues when they are attacked by harmful agents such as pathogenic bacteria and viruses; it is thought to be especially valuable as the first line of defense against these invaders. Gamma globulin is also prepared commercially to prevent or to reduce the severity of measles, hepatitis and pertussis, especially in infants and in other persons who have been exposed to the disease and are in a debilitated (weakened) condition.

Blood Studies

Many different kinds of studies may be made of the blood. Some of these have become a standard part of a routine physical examination. Machines have replaced many of the manual procedures, particularly in larger institutions.

THE HEMATOCRIT

The **hematocrit** (he-mat'o-krit) is the volume percentage of red blood cells in whole blood. It is regarded as a more reliable indicator of red cell counts than either the manual counting (using a hemocytometer) or the machine method. The hematocrit is determined by spinning a blood sample in a centrifuge for 30 minutes; in this way the cellular elements are separated out from the plasma.

The hematocrit is expressed as milliliters of blood or as volumes per 100 milliliters. Thus, in the laboratory you may read a report stating "hematocrit, 38" which would mean that the space occupied by the red blood cells was 38 percent of the whole blood volume. In males the normal range is 45 to 50 volumes per 100 ml., in females it is 40 to 45 volumes per 100 ml. Values much above or much below these point to an abnormal situation requiring further study.

BLOOD CELL COUNTS

An apparatus for counting the number of blood cells is called a **hemocytometer** (he-mo-si-tom'e-ter), illustrated in Fig. 3.4. It consists essentially of two tubes, one for red cells and the other for white cells, and an accurately ruled glass slide for viewing the blood samples under a microscope; the purpose of the gridiron rulings is to provide a fixed area in which the cells can be counted.

In most larger laboratories automatic machines such as the Coulter counter are used. If the machine breaks down or if the laboratory is small and cannot afford to purchase a machine, the hemocytometer must be used. The normal count for red cells varies from about 4.5 to 5.5 million cells per cubic millimeter (a cubic millimeter is a very tiny drop of blood). The leukocyte count varies from 5000 to 9000 per cubic millimeter.

In **leukopenia** the white count is below 5000. It is characteristic of a few infections such as malaria and measles, as well as certain disorders of the blood-forming organs.

Leukocytosis (lu-ko-si-to'sis) means that the white blood count is in excess of 9000 or 10,000 per cubic millimeter. It is particularly characteristic of most infections. It may occur also after hemorrhage and in gout and uremia, a result of kidney disease.

THE BLOOD SLIDE (OR SMEAR)

A drop of blood is spread very thinly and carefully over a glass slide. A special stain is applied to differentiate the otherwise colorless white cells, and then this slide is studied under the microscope. Abnormal red cells which are characteristic of certain anemias may be noted, and malarial or other parasites may be found. Abnormalities in the

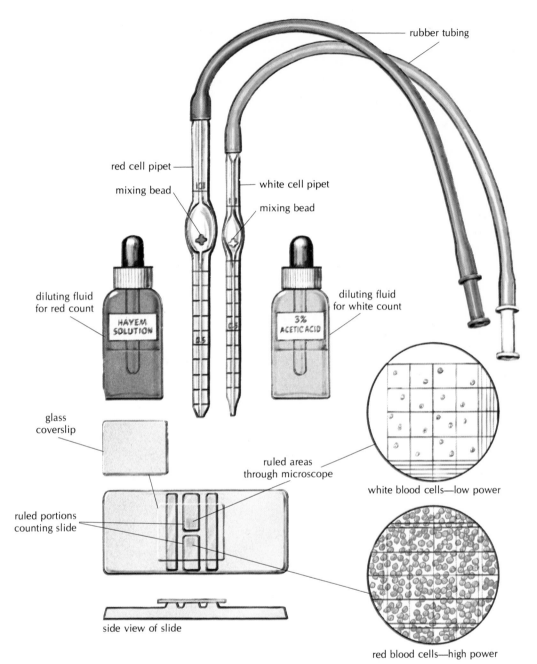

Figure 3.4. Parts of a hemocytometer.

white cells also are observed. In addition, the **differential white count** (i.e., an estimation of the percentage of each type of white cell) is done using the same stained blood slide. Such a count is an important aid to the physician in making a diagnosis.

AMOUNT OF HEMOGLOBIN

It is important to know that a person has an adequate amount of hemoglobin so that the tissues are assured a sufficient supply of oxygen. This is determined by means of a **hemometer** (he-mom′e-ter), also known as a

hemoglobinometer (he-mo-glo-bi-nom'e-ter). These devices vary in design, but in general the principle is that a comparison is made between the blood and a standard color scale. The normal amount of hemoglobin varies from about 14 to 16 grams per 100 cc. of blood. Too little hemoglobin is a factor in anemia. Because of the rather high percentage of the element iron in hemoglobin, it is important that the diet contain sufficient amounts of available iron, as found in oatmeal, eggs, molasses and animal organs (heart, kidneys and liver).

Blood Chemistry Tests

Batteries of tests on blood serum are often done by machine. One, the Sequential Multiple Analyzer (SMA), provides for the running of some 12 or more tests per minute. Tests for electrolytes such as sodium, potassium, chloride and bicarbonate, plus enzyme tests such as those for alkaline **phosphatase** (fos'fah-tase) and **transaminase** (trans-am'i-nase), may be included in this battery of tests. Others of importance include blood urea nitrogen (BUN), blood sugar, cholesterol and triglyceride evaluations. All of these may also be done manually using test tubes containing various chemicals that yield color changes or other kinds of determinants.

Many of these blood serum tests help in evaluating disorders that may involve such vital organs as the heart, kidneys, liver and pancreas. For example, the presence of more than the normal amount of glucose (sugar) dissolved in the blood is called **hyperglycemia** (hi-per-gli-se'me-ah) and is found most frequently in diabetic persons. Sometimes several evaluations of sugar content are done following the administration of a known amount of glucose. This procedure is called the **glucose tolerance test** and usually is given along with another test which determines the amount of sugar in the urine. This combination of tests can indicate faulty cell metabolism.

Summary

1. **General characteristics of blood.**
   A. Can be considered a tissue.
   B. Thick fluid of varying red color.
   C. Quantity: about 6 quarts in average adult.
2. **Main purposes of blood.**
   A. Transportation (including heat exchange).
   B. Defends body against disease.
3. **Prime elements of blood.**
   A. Plasma (liquid element).
   B. Formed elements (cell and cell-derived elements).
      (1) Erythrocytes (red cells).
      (2) Leukocytes (white cells).
      (3) Platelets (cellular fragments that cause clotting).
4. **Plasma.**
   A. Ninety percent water.
   B. Remainder: proteins, carbohydrates, lipids, salts.
5. **Formed elements (corpuscles)**—characteristics.
   A. Erythrocytes: carry oxygen; main constituent is hemoglobin, which absorbs the gases; nonnucleated; most numerous of corpuscles.
   B. Leukocytes: combat disease; nucleated; several different forms.
   C. Platelets: fragments of cells; dissolve and combine with fibrinogen to form the clot (fibrin).
6. **Origin of corpuscles.**
   A. Erythrocytes: formed in red marrow; immature cells nucleated; mature cells lose nucleus before going to blood stream.
   B. Leukocytes: formed mostly in red marrow; one form (lymphocytes) originate in various lymphoid tissues.
   C. Platelets: essential to blood clotting.
7. **Blood groups.**
   A. Not all blood types compatible.
   B. Blood types (A, B, AB, O).
   C. Mixing of incompatible bloods may result in agglutination of red cells or hemolysis.

8. **Rh factor.**
   A. Most people have special red cell protein; are Rh positive.
   B. Minority lack it; are Rh negative.
   C. Rh negative mother may have Rh positive baby (factor inherited from father). Mother's blood may cause destruction of baby's red cells.
   D. Above condition called erythroblastosis fetalis, prevented by the use of an immune globulin.
9. **Determination of parenthood:** blood grouping inherited. Blood typing can prove who is not the father and who possibly is, but not definitely who is.
10. **Blood banks:** Blood bottled and stored for emergencies. Type O most commonly used (is universal donor).
11. **Blood derivatives.**
    A. Plasma: separated, dried, stored, reactivated, used in emergencies.
    B. Serum: taken from animals, produces immunity.
    C. Others: paste of blood cells; gamma globulins.
12. **Blood studies.**
    A. Hematocrit, volume percent of red cells.
    B. Total red and white blood counts.
       (1) Apparatus used called hemocytometer, or machine such as Coulter counter.
       (2) Too few white cells is called leukopenia.
       (3) More than normal number of white cells is known as leukocytosis.
    C. Examine blood slide for parasites; used also for differential white count.
    D. Differential white count.
    E. Hemoglobin amount estimated, using hemometer.
13. **Blood chemistry tests.**
    A. Machine such as SMA can perform 12 tests per minute.
    B. Electrolytes and enzyme determinations.
    C. Glucose, urea, others for status of organ function.

Questions and Problems

1. How does the color of blood vary with the amount of oxygenation?
2. Name the 2 main purposes of the blood.
3. Name the 2 prime elements of the blood.
4. Name and describe the 3 main groups of cellular structures in blood.
5. Name 4 main ingredients of blood plasma. What are their purposes?
6. What is the main function of erythrocytes? leukocytes? platelets? Where does each originate?
7. What are the names usually given to the 4 main blood groups? What are the factors that are the reason for the different groupings?
8. What is the Rh factor? What proportion of people possess this factor? In what situations is this factor of medical importance? Why?
9. Are blood groups inherited? How is this fact made use of in paternity cases?
10. What are the advantages of blood banks? Are there any disadvantages? If so, what are they, and is there a way of counteracting these disadvantages?
11. What are some of the conditions for which blood transfusions are useful?
12. What precautions should always be taken before a transfusion is given?
13. What substances obtained from the blood may be useful in the treatment of the sick, and in what way is each of these blood derivatives used?
14. What is the value of the hematocrit?
15. What are some of the conditions that are due to abnormal white counts?
16. What is determined by the blood smear?
17. What are some evaluations made by blood chemistry tests?

Body Temperature and Its Regulation

Body Temperature and Homeostasis

In an earlier chapter it was pointed out that heat is an important by-product of the many chemical activities constantly going on in the tissues all over the body. Simultaneously, heat is always being lost through a variety of outlets. Yet, by virtue of a number of regulatory devices the body temperature remains constant within quite narrow limits under normal conditions. The maintenance of a constant temperature in spite of both internal and external influences is one phase of the concept of homeostasis, the tendency of the body processes to maintain a normal state despite forces that tend to alter them. Other examples of the maintenance of homeostasis are the heart rate, respiratory rate and blood pressure, which all tend to remain within the so-called normal limits. In addition to these more obvious types of homeostasis there are many important examples that involve the composition of body fluids.

Heat Production

Heat is produced when oxygen combines with food products in the body cells. The amount of heat produced by a given organ varies with the kind of tissue and with its activity. While at rest, muscles may produce as little as 25 percent of the total body heat, but when numbers of muscles contract, the heat production may be multiplied hundreds of times. Under basal conditions (rest) the abdominal organs, particularly the liver, produce about one half of the body heat; but during vigorous muscular activity this ratio is greatly changed. While the body is at rest, the brain may produce 15 percent of the body heat, but an increase in activity in nerve tissue produces very little increase in heat production. The largest amount of heat, therefore, is produced in the muscles and the glands. It would seem from this description that some parts of the body would tend to become much warmer than others. The circulating blood, however, distributes heat fairly evenly throughout the entire body.

The rate at which heat is produced is affected by a number of factors. When the body is at complete rest (basal condition), the glandular organs such as the liver continue to add some heat constantly with but slight variations. But the amount of heat produced in muscles during activity is hundreds of times as great as during rest. In addition to these causes of variation, certain hormones, such as thyroxine from the thyroid gland and epinephrine (adrenaline) from the medulla of the adrenal gland, may increase the rate of heat production. The intake of food also is accompanied by increased heat production. The reasons are not entirely clear. More fuel is poured into the blood and is therefore more readily available for cellular "combustion." The glandular structures and the muscles of the digestive system generate additional heat as they set to work. This does not account for all the increase, however, nor does it account for the much greater increase in metabolism following a meal containing large amounts of protein. Whatever the reasons, the intake of food definitely increases the chemical activities that go on in the body and thus adds to heat production.

Heat Loss

More than 80 percent of heat loss occurs through the skin (see Chapter 5). The remaining 15 to 20 percent is dissipated via the respiratory system and with the urine and feces. Networks of blood vessels in the deeper part (corium, or dermis) of the skin are capable of bringing considerable quantities of blood near the surface so that heat can be dissipated to the outside. This can occur in several ways. Heat can be transferred to the surrounding air (**conduction**). Heat also travels from its source in the form of heat waves or rays (**radiation**). If the air is moving so that the layer of heated air next to the body is constantly being carried away and replaced with cooler air (as by an electric fan), the process is known as **convection**. Finally, heat loss may be produced by **evaporation**. Any liquid uses heat during the process of changing to the vapor state. Rub some alcohol on your arm; it evaporates

rapidly, and in so doing uses so much heat, taking it from the skin, that your arm feels cold. Perspiration does the same thing, though not as quickly. The rate of heat loss through evaporation depends upon the humidity of the surrounding air. When this exceeds 60 percent or so, perspiration will not evaporate so readily; and one feels generally miserable unless some other means such as convection (by a fan) can be resorted to.

If the temperature of the surrounding air is lower than that of the body, excessive heat loss is prevented by both natural and artificial means. Clothing checks heat loss by trapping "dead air" both in its material and its layers. This noncirculating air is a good insulator. An effective natural insulation against cold is the layer of fat under the skin. Even though the skin temperature may be low, this fatty tissue prevents the deeper tissues from losing too much heat. This layer is on the average slightly thicker in the female than in the male. Naturally there are individual variations, but as a rule the degree of insulation depends on the thickness of this layer of subcutaneous fat.

Other factors that play a part in heat loss include the volume of tissue compared with the amount of skin surface. Just as a child loses heat more rapidly than an adult, so such parts as the fingers and the toes are affected more by exposure to cold, because in each case there is a greater amount of skin compared with total tissue volume.

Temperature Regulation

Since the body temperature remains almost constant in spite of the wide variations in the rate of heat production or loss, obviously there must be a temperature regulator. Actually, many areas of the body take part in this process, but the most important heat regulating center is a section inside the brain, located just above the pituitary gland, called the **hypothalamus** (hi-po-thal'ah-mus). Some of the cells in the hypothalamus control the production of heat in the body tissues, while another group of cells controls heat loss. This control comes about in response to the heat brought to the brain by the blood as well as to nerve impulses from the temperature receptors in the skin. If these two factors indicate that too much heat is being lost, impulses are sent quickly from the brain to the involuntary (autonomic) nervous system which in turn causes constriction of the skin blood vessels in order to reduce heat loss. Other impulses are sent to the muscles to cause shivering, a rhythmic contraction of many body muscles, which results in increased heat production. The output of epinephrine may be increased, also, if conditions call for it. The smooth muscle about the hair roots contracts, forming "gooseflesh."

If, on the other hand, there is danger of overheating, the hypothalamus will transmit impulses which stimulate the sweat glands to increased activity and also dilate the blood vessels in the skin so that there is increased blood flow with a correspondingly greater loss of heat. The hypothalamus also may encourage relaxation of muscles and thus minimize the production of heat in these organs (see Chapter 7).

Muscles are especially important in temperature regulation because variations in the amount of activity of these large masses of tissue can readily increase or decrease the total amounts of heat produced according to the needs of the body. Since muscles form roughly one third of the bulk of the body, either an involuntary or a purposeful increase in the activity of this big group of organs can form enough heat to offset considerable decrease in the temperature of the environment.

Normal Body Temperature

The normal temperature range as obtained by the usual thermometers may extend from 97° to 100°F (36.2° to 37.6°C). Temperature of the body varies with the time of day. Usually it is lower in the early morning, since the muscles have been relaxed and no food has been taken in for several hours. Temperature tends to be higher in the late afternoon and evening because of physical activity and consumption of food.

Normal temperature also varies with the part of the body. Skin temperature as obtained in the armpit, **axilla** (ak-sil'ah), is lower than mouth temperature, and mouth temperature is a degree or so lower than rectal temperature. If it were possible to place a thermometer inside the liver, it is believed that it would register a degree or more higher than the rectal temperature. The temperature within a muscle might be even higher during its activity.

Although the Fahrenheit scale is used in the United States, in most parts of the world temperature is measured using the **Celsius** (sel'se-us) thermometer. The ice point is at 0 and the normal boiling point of water is at 100°, the interval between these two points being divided into 100 equal units. The Celsius scale is also called the centigrade scale (think of 100 cents).

On the Fahrenheit thermometer, there are 180 divisions; one Fahrenheit division equal 5/9 Celsius division, and one Celsius division equals 9/5 Fahrenheit divisions. It is a simple matter to convert one to the other: C = (F − 32) × 5/9; F = (C × 9/5) + 32.

1. 0°C = 32°F = The freezing point of water.
2. 10°C = 50°F = Cool, mild weather.
3. 20°C = 60°F = Comfortable weather temperature.
4. 30°C = 86°F = Warm weather.
5. 37°C = 98.6°F = Normal body temperature.
6. 40°C = 104°F = Hot weather.
7. 100°C = 212°F = Boiling temperature of water.

Abnormal Body Temperature

FEVER

Fever is a condition in which the body temperature is higher than normal. Usually the presence of fever is due to an infection, though there can be many other causes such as malignancies, brain injuries, toxic reactions, reactions to vaccines, and diseases involving the central nervous system. Sometimes emotional bouts can bring on a fever.

Curiously enough, fever usually is preceded by a chill—that is, a violent attack of shivering and a sensation of cold that such measures as blankets and hot water bottles seem unable to relieve. At the same time heat is being generated and stored in the body; and when the chill subsides, the body temperature is elevated.

The old adage that a fever should be starved is completely wrong. During a fever there is an increase in metabolism that is usually proportional to the amount of fever. In addition to the use of available sugar and fat there is an increase in the use of protein, and during the first week or so of a fever there is definite evidence of destruction of body protein. A high calorie diet with plenty of protein is therefore desirable.

When a fever ends, sometimes the drop in temperature to normal occurs very rapidly. This sudden fall in temperature is called the **crisis**, and is usually accompanied by symptoms indicating rapid heat loss: profuse perspiration, muscular relaxation and dilated blood vessels in the skin. A gradual drop in temperature, on the other hand, is known as **lysis**.

The mechanism of fever production is not completely understood, but we might think of the hypothalamus as a thermostat which is set higher at this time. This change in the heat-regulating mechanism often follows the injection of a foreign protein or the entrance into the blood stream of bacteria or their toxins. Up to a point fever may be beneficial because it steps up **phagocytosis** (fag-o-si-to′sis), the process by which white blood cells surround, engulf and digest bacteria and other foreign bodies; inhibits the growth of certain organisms (such as the spirochete of syphilis); and probably encourages the production of antibodies. Sometimes fever is induced on purpose as a form of treatment, particularly for syphilis of the nervous system.

The body cannot endure temperatures beyond about the level of 112°F (about 45°C) because at that point tissues are irreversibly damaged and death occurs.

EFFECTS OF EXTREME
OUTSIDE TEMPERATURES

The body's heat-regulating devices are efficient, but there is a limit to what they can accomplish. If the outside temperature is too high, one may perspire so much that dehydration and **heat exhaustion** can result. Heat exhaustion is a condition caused by excessive salt loss, and its symptoms include muscle cramps, dizziness, vomiting and fainting. This condition usually can be forestalled by taking salt tablets in hot weather.

Sunstroke (sometimes called **heat stroke**) also is caused by high outside temperatures. It differs from heat exhaustion in that one of the heat regulators is affected; namely, the sweat glands. Dehydration begins a chain of events which terminates in decreased blood supply to the skin and diminished secretion of perspiration. As a consequence, the body temperature rockets up to a level that can be fatal. The victim of sunstroke exhibits many of the symptoms of heat exhaustion (i.e., dizziness, fainting) but with this significant difference: there is an absence of perspiration; the skin is dry and flushed. Sunstroke is an extremely serious emergency. The most important first aid measure is to lower the temperature; otherwise permanent brain damage can result. Cooling of the body is accomplished by immersing the victim in cool water or else by spraying him with it. Ice should be applied to the head, and cold drinks administered if the patient is conscious.

The body is no more capable of coping with prolonged exposure to cold than to heat. If, for example, the body is immersed in cold water for a time, the water (a better heat conductor than air) removes more heat from the body than can be replaced, and the body temperature falls. This can happen too, of course, in cold air—particularly when clothing is inadequate. An excessively low body temperature is termed **hypothermia** (hi-po-ther′me-ah), and its main effects are lowered respiratory rate and blood pressure and a feeling of drowsiness finally ending in coma and perhaps death. Hypothermia, the opposite of fever, is caused mainly by prolonged exposure to cold, rarely by abnormal conditions within the body.

Summary

1. **Body temperature and homeostasis.**
   A. Homeostasis: tendency of body processes to maintain normal state.
   B. Body temperature regulation a phase of homeostasis.
2. **Heat production.**
   A. Produced constantly in metabolic processes.
   B. Muscles and glands produce most heat.
   C. Heat distributed through body by blood.
   D. Heat production rate determined by muscle and glandular activity, food intake, external temperature.
3. **Heat loss.**
   A. Outlets: chiefly skin; also respiratory system, urine, feces.
   B. Heat loss from skin: conduction, radiation, convection, evaporation.
   C. Excess heat loss prevented by artificial means (clothing), natural means (subcutaneous fat).
4. **Temperature regulation.**
   A. Chief center is hypothalamus.
   B. Information reaches it via blood and also nerves from temperature receptors.
   C. Hypothalamus causes either increased heat production (blood vessel constriction in skin, shivering, "gooseflesh," possibly increased epinephrine) or increased heat loss (activation of sweat glands, blood vessel dilation, muscular relaxation).
5. **Normal body temperature.**
   A. Normal range 97°–100°F (36.2° to 37.6°C).
   B. Varies with time of day, part of body.
   C. On Celsius, or centigrade, scale, water freezes at 0° and boils at 100°; on Fahrenheit scale water freezes at 32° and boils at 212°.
6. **Abnormal body temperature.**
   A. Fever.
      (1) Abnormally high body temperature, usually caused by infection.
      (2) Preceded by chill; terminates in crisis or lysis.
      (3) Causes destruction of some body protein; increases phagocytosis and possibly antibody production.
   B. Extremes in outside temperature.
      (1) Heat: can cause heat exhaustion or sunstroke.
      (2) Cold: can cause hypothermia (sometimes but rarely of pathological origin); cold injuries.

Questions and Problems

1. What is homeostasis? Name 4 aspects of it.
2. How is heat produced in the body? What structures produce the most heat during increased activity?
3. Name 4 factors affecting heat production.
4. By what channels is heat lost from the body?
5. Name 4 ways in which heat escapes to the environment.
6. In what ways is heat kept in the body?
7. Name the main temperature regulator and describe what it does when the body is too hot and when it is too cold. What part do muscles play?
8. What is the normal body temperature range? How does it vary with respect to the time of day and the part of the body?
9. Define fever, name some aspects of its course, and list some of its beneficial and detrimental effects.
10. Name and describe 2 consequences of excessive outside heat. Why do these conditions occur?
11. What is the prime emergency measure for sunstroke?
12. What is hypothermia? Under what circumstances does it usually occur?

CHAPTER FIVE

What is the skin? •
How is the skin constructed? •
Important functions of the skin •
What can the skin tell you? •
Of what value are skin secretions? •
What are skin appendages? •

The Skin

Is the skin merely a body covering, is it an organ or is it a composite of parts that make it properly a system? Actually, it has some properties of each of these, and so may be classified in three different ways, namely:

1. It may be called an **enveloping membrane** because it is a rather thin layer of tissue covering the entire body.
2. It may be referred to as an **organ** (the largest one, in fact) because it contains several kinds of tissue, including epithelial, connective and nerve tissues.
3. The skin is also known as the **integumentary system**, "integument" (in-teg'u-ment) meaning "covering," because it includes sweat and oil glands as well as other parts that work together as a body system.

Structure of the Skin

LAYERS ENVELOPING THE BODY

The surface of the body is covered by three main layers of tissue, each different from the other in structure and function (Fig. 5.1). These are:

1. The **epidermis** (ep-i-der'mis), or outermost layer, which is subdivided into strata (stra'tah), or layers, and is made entirely of epithelial cells with no blood vessels.

2. The **dermis**, or true skin, which has a framework of connective tissue, and contains many blood vessels, nerve endings and glands.
3. The **subcutaneous** (sub-ku-ta'ne-us), or under-the-skin layer, which is a combination of elastic and fibrous tissue as well as deposits of fat (adipose tissue). This layer sometimes is referred to as the **superficial fascia** (see Chapter 2) and actually is a means of connecting the skin proper to the surface muscles. The fat in this sheet of tissue serves as insulation as well as a reserve store for energy.

THE EPIDERMIS

The outer cells of the epidermis are flat and horny. They are constantly being shed, and additional cells are pushed outward from the deeper layers of the epidermis. Since there are no blood vessels in this part of the skin, nutrition and fluids reach all the epidermal cells by means of the tissue fluid. This fluid seeps or filters out of the blood vessels in the deeper dermis and flows slowly toward the surface between the layers of skin cells. The epidermis undergoes constant change as the germinating cells of the deeper strata produce new daughter cells. The pigment granules called **melanin** (mel'ah-nin), which give the skin its color, are found in the germinating part of the epidermis. The ridges in the epidermis can be

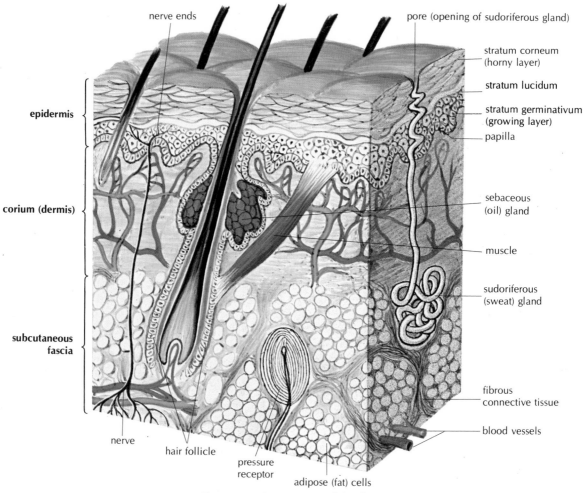

Figure 5.1. Cross section of the skin.

seen in fingerprints. These are due to elevations and depressions in the epidermis and the dermis. The deep surface of the epidermis is accurately molded upon the outer part of the dermis, which has raised and depressed areas.

THE DERMIS, OR CORIUM

The **dermis**, or **corium** (ko're-um), the so-called true skin, has a framework of elastic connective tissue and is well supplied with blood vessels and nerves. Involuntary muscle fibers also are found in the dermis, particularly where there are hairs. The thickness of the dermis as well as that of the epidermis varies so that some areas such

as the soles of the feet and the palms of the hands are covered with very thick layers of skin, while the skin of the eyelids is very thin and delicate. Most of the appendages of the skin, including the sweat and oil glands, the nails and the hairs, extend into the dermis and often deeper into the subcutaneous layer.

GLANDS OF THE SKIN

The sweat, or **sudoriferous** (su-dor-if'er-us), **glands** are coiled tubelike structures located mainly in the dermis, though sometimes in the subcutaneous tissue. Each gland has an excretory tube that extends to the surface and opens at a pore. Contrary to a popular notion, these pores do not open and close like a mouth, since there is no muscle tissue connected with them. The wax, or **ceruminous** (se-ru'min-us), **glands** in the ear canal and the **ciliary** (sil'e-er-e) **glands** on the edges of the eyelids are modifications of sweat glands.

The **sebaceous** (se-ba'shus), or **oil, glands** are saclike or alveolar (al-ve'o-lar) in structure, and their oil secretion helps to keep the hair from becoming brittle. Their ducts open most often into the hair follicles, but in some instances they open onto the skin surface. Before the baby is born, these glands produce a covering like cream cheese. This secretion is called the **vernix caseosa** (ver'niks ka-se-o'sah).

Functions of the Skin

Although the skin has several functions, the three which are by far the most important are:

1. Protection of deeper tissues against drying and against invasion by pathogenic organisms or their toxins.
2. Regulation of body temperature by dissipating heat to the surrounding air.
3. Obtaining information about the environment by means of the nerve endings which are so profusely distributed in the skin.

The intact skin is incapable of defense against sharp objects, but it is an able defender against pathogens, toxins and the process of evaporation. A break in the continuity of the skin by **trauma** (traw'mah), that is, a wound or injury of any kind, may be followed by serious infection. The care of wounds involves to a large extent the prevention of the entrance of pathogens and toxins into the deeper tissues and body fluids.

The regulation of body temperature is also a very important function of the skin. The normal temperature may vary slightly, but we think of 98.6°F. (37°C.) as the standard when a thermometer is placed in the mouth for 3 to 5 minutes. The body temperature reading may be expected to be somewhat less if taken in the axilla, and somewhat more if taken in the rectum. The skin forms a large surface for radiating body heat to the air. When the blood vessels enlarge (dilate), more blood is brought to the surface so that heat can be dissipated into the air. The evaporation of sweat from the surface of the body also helps to cool the body. As is the case with so many body functions, the matter of temperature regulation is complex and involves several parts of the body, including certain centers in the brain.

A child loses heat faster than does an adult since a higher proportion of the body is skin surface, and thus relatively more area is exposed. Therefore, it is important to prevent undue exposure to the elements in the case of infants and small children. The elderly do not produce heat in the body so easily nor to so great an extent; therefore they also should be protected against cold.

A most important function of the skin is in obtaining information from the environment. Because of the many receptors (nerve endings) for pain, touch, pressure and temperature, which are located mostly in the dermis, the skin may be regarded as one of the chief sensory organs of the body. Many of the reflexes which make it possible for the human being to adjust himself to the environment begin as sensory impulses from the skin. Here, too, the skin works in cooperation with the brain and the spinal cord to make these important functions possible.

The functions of absorption and excretion are minimal as far as the skin is concerned. The absorbing power of the skin is very limited. Excretion by the skin is limited to water and salt in perspiration; and from the point of view of removing waste products from the body the skin's excretory function is also negligible.

The human skin does not "breathe." The pores of the epidermis serve only as outlets for perspiration and oil from the glands. The normal skin secretions (oil and perspiration) are slightly acid. These secretions tend to destroy or at least to inhibit bacterial growth. This activity might be compared to that of stomach acid, which aids in destroying harmful bacteria ingested with food.

Observation of the Skin

What can the skin tell you? What do its color, texture and other attributes indicate? Much can be learned by the astute observer.

The color of the skin is dependent upon a number of factors, including:

1. The amount of pigment in the epidermis.
2. The quantity of blood circulating in the surface blood vessels.
3. The concentration of hemoglobin in the blood.
4. The presence or absence of oxygen in the blood.
5. The existence of such substances as bile, silver compounds or other chemicals in the blood.

PIGMENT OF THE SKIN

The pigment of the skin, as we have noted, is called melanin. It is found also in the hair, the middle coat of the eyeball, the iris of the eye and in certain tumors. Melanin is common to all races, but the darker peoples have a much larger quantity of it distributed in these tissues. A normal increase in this skin pigment occurs as a result of exposure to the sun. Abnormal increases in the quantity of melanin may occur either in localized areas or over the entire body surface. Diffuse spots of pigmentation may be characteristic of some endocrine disorders.

Appendages of the Skin

The nails are thin, horny plates made of translucent (i.e., partly transparent) cells that originate from the outer part of the epidermis. They are tough firm structures that aid in the protection of the distal ends of the fingers and toes.

Hair is distributed in varying amounts over most of the body surface. The part of the hair that is below the surface is the root, while the part extending outside is the shaft. The hair originates from epithelial cells that line the hair follicle which contains the hair root. The follicle is a tiny pitlike depression that extends into the deeper dermis and the subcutaneous tissue.

The glands of the skin (sebaceous and sudoriferous) discussed previously are sometimes regarded as skin appendages also.

Summary

1. **Ways in which the skin can be classified.**
   A. An enveloping membrane.
   B. The largest organ.
   C. A body system (integumentary).
2. **Structure of the skin.**
   A. Layers.
   (1) Epidermis.
   (a) Outermost layer of skin, stratified.
   (b) Made of epithelial cells with no blood vessels.
   (c) Undergoes constant cellular change.
   (d) Contains pigment (melanin).
   (2) Dermis.
   (a) True skin, connective tissue framework.
   (b) Contains blood vessels, nerves and glands.
   (3) Subcutaneous layer.
   (a) Called superficial fascia.
   (b) Fat deposits are a reserve energy store.
   B. Glands of the skin (in dermis).
   (1) Sudoriferous: sweat glands.
   (2) Sebaceous: oil glands.
   (3) Ciliary and ceruminous: modified sweat glands.
3. **Functions of the skin.**
   A. Protection against pathogens, toxins and drying of under tissues.
   B. Regulation of body temperature.
   C. Sensory organ for pain, touch, heat, cold and pressure.
   D. Excretory function: limited to water and salt.
4. **Observation of the skin.**
   A. Pigmentation (melanin content).
   (1) Melanin content varies with race.
   (2) Sunlight causes increased pigmentation.
   (3) Abnormal melanin content or distribution can be a sign of disease.
5. **Appendages of the skin, nails and hair.**

Questions and Problems

1. What characteristics of the skin classify it as a membrane, as an organ, as a system? What term do you think fits it best and why?

2. Of what type of cells is the epidermis composed?

3. Explain how nutrients and oxygen reach the cells of the epidermis.

4. What kind of tissue forms the framework of the dermis?

5. What glands are found in the dermis, and how do their secretions reach the surface? What is the function of the skin pores?

6. What is the subcutaneous tissue made of, and what is another name for it?

7. Explain the 3 most important functions of the skin.

8. What are the facts about the skin as an organ of respiration and excretion?

9. What are the most important contributors to the color of the skin, normally?

10. What is the value of the acid secretions of the skin?

11. Describe the origin and structure of the nails.

12. What and where is the hair shaft? Describe the hair follicle and its relationship to the hair root.

Bones, Joints and Muscles

The Musculoskeletal System

The bones are the framework around which the body is constructed. The muscles might be considered its motive power. The joints, where the bones come together, allow the bones, powered by the muscles, a great variety and range of motion. The combination of bones, joints, muscles and related connective tissues is known as the **musculoskeletal system**.

The Bones

BONE STRUCTURE

In an earlier chapter we saw that the bones are composed chiefly of bone tissue, known as **osseous** (os'e-us) tissue. It should be understood at the outset that a bone is anything but lifeless. Even though the spaces between the cells of bone tissue are permeated with stony deposits of calcium,

these cells themselves are very much alive. Bones are organs, with their own system of blood and lymphatic vessels and nerves.

In the embryo (the early development stage of a baby) most of the bones-to-be are cartilage. During the second and third months following conception calcium compounds begin to be deposited in the skeleton. The process of depositing these hard calcium salts continues throughout life, more rapidly at some times than at others. The bones of small children are relatively pliable because they contain a larger proportion of cartilage and a smaller amount of the firm calcium salts than those of adults. In the elderly person there is much less of the softer tissues such as cartilage and a high proportion of calcium salts; therefore, the bones are brittle. Fractures of bones in older people heal with difficulty mainly because of this relatively high proportion of inert material and the small amount of the more vascular softer tissues.

There are two kinds of marrow: **red marrow**, found in certain parts of all bones, which manufactures most of the blood cells; and **yellow marrow** of the "soup bone" type, found chiefly in the central cavities of the long bones. Yellow marrow is largely fat.

Bones are covered on the outside (except at the joint region) by a membrane called **periosteum**. The central cavity of the long bones is lined with a thinner membrane known as **endosteum** (en-dos'te-um).

These bone membranes contain blood and lymph vessels as well as nerves—the latter making their presence felt if, for instance, one suffers a blow on the shin. The chief functions of endosteum and periosteum are to produce bone during the period of the individual's growth, and to aid in repair following an injury such as a fracture.

MAIN FUNCTIONS OF BONES

Bones have a number of functions, many of which are not at all obvious. Some of these are:

1. To serve as a firm framework for the entire body.
2. To protect such delicate structures as the brain and the spinal cord.
3. To serve as levers, which are actuated by the muscles that are attached to them.
4. To serve as a storehouse for calcium, which may be removed to become a part of the blood if there is not enough calcium in the diet.
5. To produce blood cells.

DIVISIONS OF THE SKELETON

The complete bony framework of the body is known as the **skeleton**, and it may be divided into two main groups of bones (Fig. 6.1):

1. The **axial** (ak'se-al) **skeleton**, which includes the bony framework of the head and the trunk.
2. The **appendicular** (ap-en-dik'u-lar) **skeleton**, which forms the framework for those parts usually referred to as the arms and legs, but called the **extremities** by the biologists.

The framework of the head

The bony framework of the head is called the skull, and it is subdivided into two parts, namely (Figs. 6.2-6.4):

1. The **cranium**, which is a rounded box that encloses the brain and is made of eight distinct cranial bones.
2. The **facial portion** of the skull, composed of 14 separate bones.

The bones that form the cranium are:

1. The **frontal bone**, which forms the framework for the forehead, the roof between the eyeballs and the frontal parts of the cerebrum, and contains two air spaces (sinuses).
2. The two **parietal bones**, which form the larger part of the upper and side walls of the cranium.
3. The two **temporal bones**, which form the lower sides and part of the base of the central areas of the skull, and which contain the mastoid sinuses as well as the parts of the ear.
4. The single **ethmoid** (eth'moid), which is a delicate, spongy bone located between the eyes and forming a part of the cranial floor between the frontal lobes of the brain and the upper nasal cavities.
5. The **sphenoid** (sfe'noid) **bone**, which is a bat-shaped bone that extends behind the eyes and forms a part of the base of the skull in this region.
6. The single **occipital** (ok-sip'i-tal) **bone** which is located at the back of the skull and includes most of the base of the skull. It extends at the base to meet with the sphenoid medially and the temporal and parietal at the sides.

The places at which the cranial bones join are called **sutures**.

The facial bones include:

1. The **mandible** or lower jaw bone, which is the only movable bone of the skull.
2. The **maxillae** (mak-sil'e), which form the upper jaw. Each maxilla contains a rather large air space called the **maxillary sinus**.
3. The **zygomatic** (zi-go-mat'ik) **bones**, one on each side, which form the higher portion of the cheek.

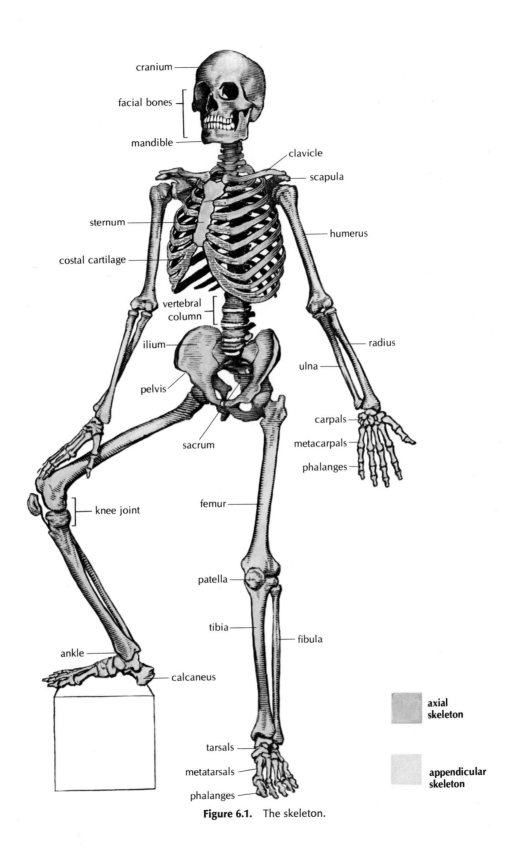

cranium

facial bones

mandible

clavicle

scapula

sternum

humerus

costal cartilage

vertebral column

radius

ilium

ulna

pelvis

carpals

metacarpals

sacrum

phalanges

knee joint

femur

patella

tibia

fibula

ankle

calcaneus

tarsals

metatarsals

phalanges

axial skeleton

appendicular skeleton

Figure 6.1. The skeleton.

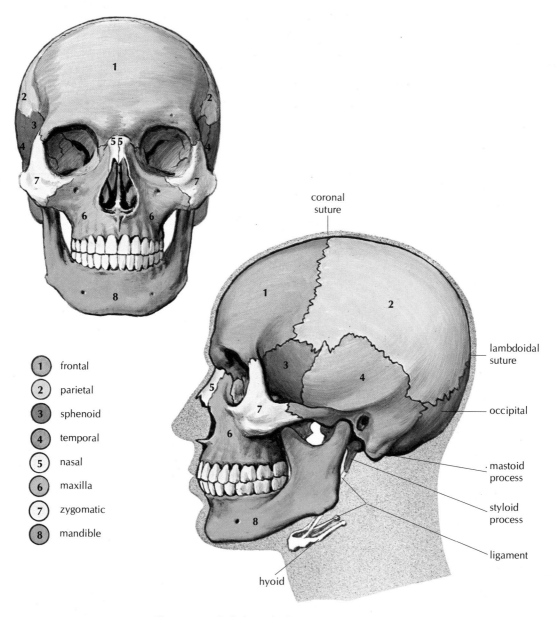

1 frontal
2 parietal
3 sphenoid
4 temporal
5 nasal
6 maxilla
7 zygomatic
8 mandible

Figure 6.2. Skull, from the front and from the left.

4. Smaller bones of the face, which include the two **lacrimal bones**, one at the corner of each eye; the single **vomer** (vo'mer) that forms the lower part of the nasal septum; the paired **palatine bones** at the back of the hard palate; the slender **nasal bones**, supporting the bridge of the nose; and the paired inferior **nasal conchae** (kong'ke). The paired superior and middle nasal conchae are all part of the ethmoid, one of the cranial bones.

In addition to the bones of the cranium and the facial bones there are six tiny bones or **ossicles** (three pairs) in the middle ear (Chapter 8) and a single horseshoe- or U-shaped bone that lies just below the skull proper, called the **hyoid** (hi'oid) **bone**, to which the tongue is attached. Actually, the hyoid bone may be thought of as forming

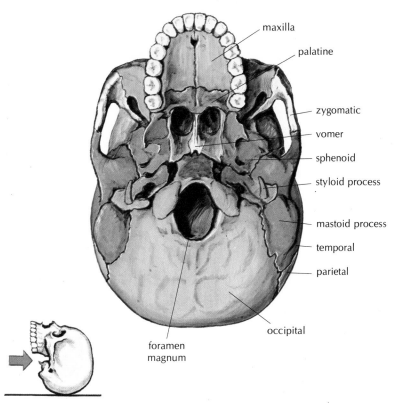

maxilla

palatine

zygomatic

vomer

sphenoid

styloid process

mastoid process

temporal

parietal

occipital

foramen magnum

Figure 6.3. Skull from below, lower jaw removed.

the boundary between some of the head structures and those of the neck. It is suspended by a ligament on each side from the temporal bones and can be felt near the mandible or lower jaw bone in the upper neck region.

Openings in the base of the skull provide spaces for the entrance and exit of many blood vessels, nerves and other structures. Projections and slightly elevated portions of the bones provide for the attachment of muscles and sometimes contain delicate structures, as, for example, the part that encloses the middle and internal parts of the ear. Air spaces called sinuses provide lightness and serve as resonating chambers for the voice.

The framework of the trunk

The bones of the trunk include the **vertebral** (ver'te-bral) **column** and **rib cage**. The vertebral column is made of a series of irregularly shaped bones, numbering 33 or 34 in the child; but because of unions that occur later in the lower part of the spine,

there usually are 26 separate bones in the adult column (Figs. 6.5 and 6.6). Each of these **vertebrae** (ver'te-bre) has a body at the front, an opening for the spinal cord toward the back, and a group of projections extending from the arch that encircles the canal, or opening, for the spinal cord. Between the vertebrae are plates of cartilage that absorb shocks and make for flexibility. The spinal column is divided into five regions, as follows:

1. The **cervical** (ser'vi-kal) section, made of seven vertebrae which form the main framework of the neck.
2. The **thoracic** region, which contains 12 bones, and has a distinct outward or convex curve. It differs from all other parts of the column in that here are attached 12 pairs of ribs.
3. The **lumbar** area, which contains five somewhat larger vertebrae than are found in the first two sections. It is usually curved inward (concave).
4. The **sacral** (sa'kral) portion of the column, which is made of five vertebrae in the child, but which under-

59

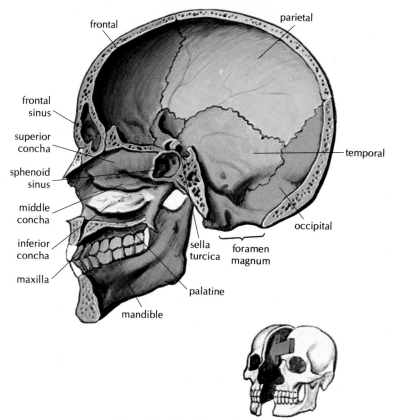

frontal

parietal

frontal
sinus

superior
concha

sphenoid
sinus

middle
concha

inferior
concha

maxilla

mandible

temporal

occipital

sella
turcica

foramen
magnum

palatine

Figure 6.4. Skull, internal view.

goes a union of bones so that the adult has one **sacrum** (sa′krum). This serves to complete the framework of the pelvic girdle at the back.

5. The **coccygeal** (kok-sij′e-al) or tail part, made of four or five bones in the child, but which fuses to form a single coccyx in the adult.

Primary and secondary curves

In the infant the entire vertebral column is concave ventrally, forming the primary curve. As the child learns to walk and then run, secondary curves appear in the cervical and lumbar regions. They are convex ventrally, concave dorsally.

The rib cage

The rib cage is made up of 24 ribs attached to the vertebral column at the back (dorsally) and mostly to the breast bone or **sternum** (ster′num) in front. It serves to

support the chest and protect the heart, the lungs and other organs. Each rib is a slender, somewhat flattened bone that curves and is slightly twisted. The variation in the anterior attachment to the sternum has led to the following grouping of the ribs, or **costae** (kos′te):

1. **True ribs**, which are the first seven pairs. These are attached directly to the sternum by means of individual extensions called **costal** (kos′tal) **cartilages.**

2. **False costae**, which include the remaining five pairs of ribs. Of these, the eighth, ninth and tenth pairs attach to the cartilage of the rib above. The final two pairs do not attach to any bones at all in front and are called **floating ribs.** These last two ribs are relatively short and so do not extend around to the front of the body.

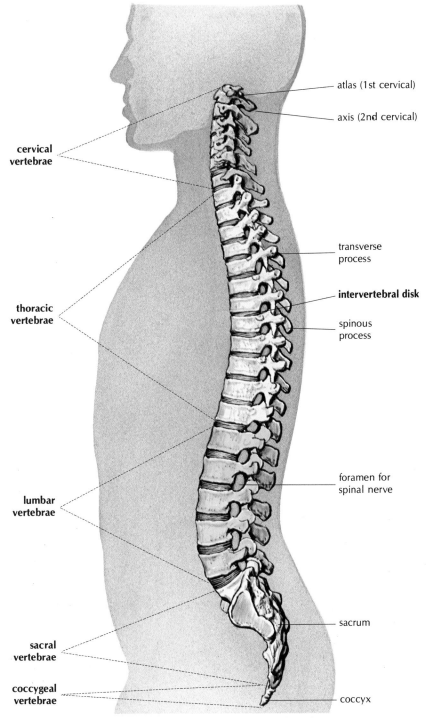

atlas (1st cervical)

axis (2nd cervical)

cervical
vertebrae

transverse
process

intervertebral disk

thoracic
vertebrae

spinous
process

foramen for
spinal nerve

lumbar
vertebrae

sacrum

sacral
vertebrae

coccygeal
vertebrae

coccyx

Figure 6.5. Vertebral column from the side.

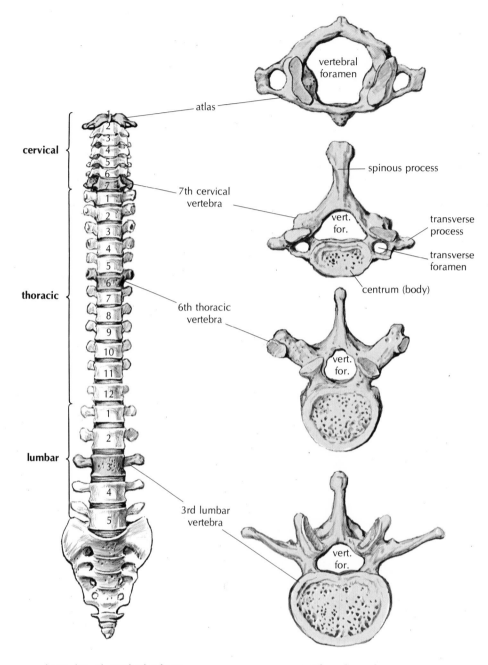

cervical

1
atlas
vertebral
foramen

2
3
4
5
6
7

7th cervical
vertebra

spinous process

vert.
for.

transverse
process

transverse
foramen

centrum (body)

thoracic

1
2
3
4
5
6
7
8
9
10
11
12

6th thoracic
vertebra

vert.
for.

lumbar

1
2
3
4
5

3rd lumbar
vertebra

vert.
for.

front view of vertebral column   vertebrae from above

Figure 6.6.   Front view of vertebral column; vertebrae from above.

The bones of the extremities

The framework of the extremities includes the longest bones in the body. Long bones contain a special central canal not found in any other bones. This **medullary** (med'u-lar-e) **canal** or marrow cavity is filled with soft tissue, largely fat, which gives the bone lightness and provides a reservoir for storage of fuel in case of need. This bone structure can be compared with a bamboo stick in its relative lightness and strength. Other bones of the extremities are flat, irregular, or short; and like bones elsewhere in the body, contain red marrow.

The extremities are considered as having two divisions: upper and lower. The upper extremities include the shoulders, the arms (between the shoulders and the elbows), the forearms (between the elbows and the wrists), the wrists, the hands and the fingers. The lower extremities include the hips (pelvic girdle), the thighs (between the hips and the knees), the legs (between the knees and the ankles), the ankles, the feet and the toes.

The bones of the upper extremities are grouped as follows:

1. The shoulder or **pectoral** (pek'to-ral) **girdle**, which contains the collar bone, or **clavicle** (klav'i-kle), and the shoulder blade, or **scapula** (skap'u-lah).

2. The bone of the arm proper, the **humerus** (hu'mer-us), which connects at the top with the scapula and with the two forearm bones at the elbow.

3. The forearm bones, which are the **ulna** (ul'nah) on the same side as the little finger (medial), and the **radius** (ra'de-us) on the thumb side, which is lateral when the palm is directed forward.

4. The **carpal** (kar'pal) **bones** of the wrist, which number eight for each side, all different and separately named.

5. The **metacarpal bones** of the palm of the hand, which are five in number on either side. Their distal ends form the knuckles.

6. The **phalanges** (fa-lan'jez), which are the bones forming the framework of the fingers, numbering 14 for either side. Each of these bones is called a **phalanx** (fa'lanks), and exact identification may be made by using the words "distal" for the tips, "middle" for the next group, and "proximal" for those connected with the metacarpal bones. Each finger has three phalanges; the thumb has but two.

The bones of the lower extremities are grouped in a similar fashion:

1. The **pelvic girdle** of the hip region, which includes an **os coxae** (kok'se) on each side, articulating (joining) with the sacrum at the back. The pelvis proper is the lower part of the body cavity, and its boundaries comprise the inner edges of the two ossa (os'ah) coxae, the sacrum and the coccyx. The pelvis of the female is broader and lighter than that of the male, and has a greater capacity, all of which differences are an adaptation to pregnancy and childbirth (Fig. 6.7).

2. The thigh bone, called the **femur** (fe'mer). This is the longest bone in the body.

3. The **patella** (pah-tel'lah), commonly called the kneecap. This is an example of a **sesamoid** (ses'ah-moid) **bone**, a type usually encased in connective tissue cords (tendons) and designed to minimize friction. The patella is the largest sesamoid bone in the body; smaller examples of this type are found in both the upper and the lower extremities.

4. The two leg bones, which include the larger, weight-bearing bone called the **tibia** (shin bone) and the slender **fibula** (fib'u-lah), which does not extend up as high as the knee joint and is not weight-bearing.

5. Ankle and foot bones, which include the seven **tarsal bones** (of the ankle) and the five **metatarsal** (or instep) **bones**, on either side. The heel bone, or **calcaneus** (kal-ka'ne-us), is the largest of these.

6. The **phalanges** of the toes, which are counterparts of those in the fingers. There are three of these in each toe except for the great toe, which has but two.

LANDMARKS OF BONES

The contour of bones resembles the topography of an interesting and varied landscape with its hills and valleys. The projections often serve as regions for muscle attachments. There are hundreds of these prominences or **processes** with different names. Some are very important points of reference, as for example:

1. The **olecranon** (o-lek'rah-non), which

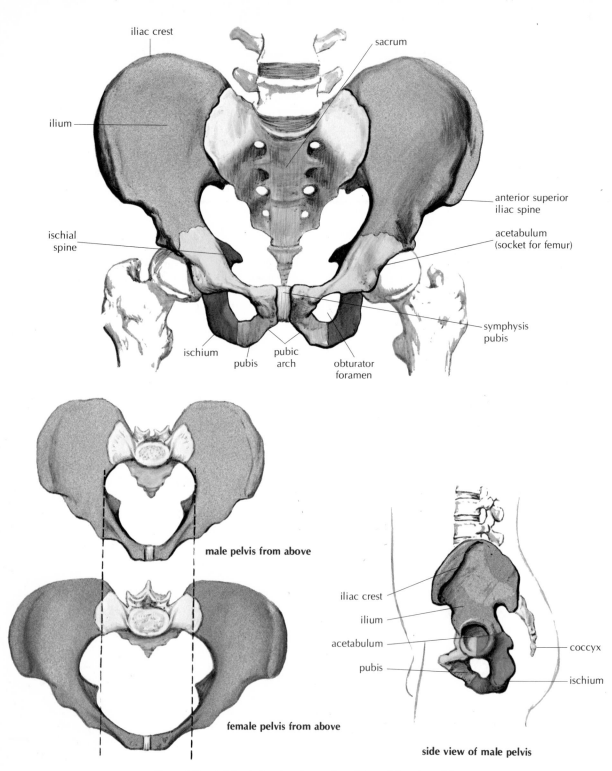

Figure 6.7. Pelvic girdle, showing male pelvis and female pelvis.

Labels in figure:

iliac crest

sacrum

ilium

anterior superior iliac spine

acetabulum (socket for femur)

ischial spine

symphysis pubis

ischium

pubis

pubic arch

obturator foramen

male pelvis from above

female pelvis from above

iliac crest

ilium

acetabulum

pubis

coccyx

ischium

side view of male pelvis

is the upper part of the ulna and forms the point of the elbow.

2. The **iliac** (il'e-ak) **spine**, which refers particularly to the pointed upper front part of the hip bone, or os coxae. There actually are three other iliac spines on each side, but this is the most important of these and is often used as a landmark, or reference point, in diagnosis and treatment.
3. The **ischial** (is'ke-al) **spine** at the back of the pelvic outlet, which is used as a point of reference during childbirth to indicate the progress of the presenting part (usually the baby's head) down the birth canal.
4. The greater **trochanters** (tro-kan'ters), which are large rounded projections located at the upper and lateral (side) portions of the femur.

Some bones are divided into areas based on the parts that were separated from each other by cartilage during childhood. An important example of this regional division is found in the pelvic girdle where the os coxae is divided into three areas:

1. The **ilium** (il'e-um), which forms the upper wing-shaped part on each side.
2. The **ischium**, which includes a part on which we sit called the **ischial tuberosity** as well as the spine mentioned in the previous paragraph.
3. The **pubis**, which forms the front portion of the os coxae and includes the region of union of the two ossa coxae in the center front to form the joint known as the **symphysis** (sim'fi-sis) **pubis**.

The skull of the infant has areas in which the bone formation is incomplete, leaving so-called soft spots. Although there are a number of these, the largest and best known is near the front at the junction of the two parietal bones with the frontal bone. It is called the **anterior fontanel** (fon-tah-nel'), and it does not usually close until the child is about 18 months old (see Fig. 6.8).

Holes that extend into or through bones are called **foramina** (fo-ram'i-nah). Numerous foramina permit the passage of blood vessels to and from the bone tissue and the marrow cavities. Larger foramina in the base of the skull and in other locations allow for the passage of cranial nerves, blood vessels, and other structures that connect with the brain. The largest such opening in the skull is called the **foramen** (fo-ra'men) **magnum**, which contains the spinal cord and related parts. The largest foramina in the entire body are found in the pelvic girdle, near the front of each os coxae, one on each side of the symphysis pubis. These are called the **obturator** (ob'tu-ra-tor) **foramina**, and are partially covered by a membrane.

Valleylike depressions on a bone surface are called **fossae** (fos'se), the singular form being **fossa** (fos'sah). Some of these are filled with muscle tissue, as is the case with the large fossae of the two scapulae. Other depressions are narrow elongated areas called **grooves**. They may allow for the passage of blood vessels or nerves as in the case of the ribs, where grooves contain intercostal nerves and vessels.

Joints

KINDS OF JOINTS

Joints, or more scientifically, **articulations**, may be defined as the region of union of two or more bones.

Joints are classified into three main groups on the basis of the degree of movement permitted:

1. **Synarthroses** (sin-ar-thro'sez), the immovable joints.
2. **Amphiarthroses** (am-fe-ar-thro'sez), the slightly movable joints.
3. **Diarthroses** (di-ar-thro'sez), the freely movable joints.

JOINT STRUCTURE

Connective tissue bands, called **ligaments**, hold the bones together and are found in connection with all the freely movable joints and many of the less movable articulations (Fig. 6.9). In some cases these completely enclose the joint and are called **capsular** (kap'su-lar) **ligaments**. Additional ligaments reinforce and help to stabilize the joints at various points. The contacting surfaces of each joint are covered by a smooth layer of gristle called the **articular** (ar-tik'u-

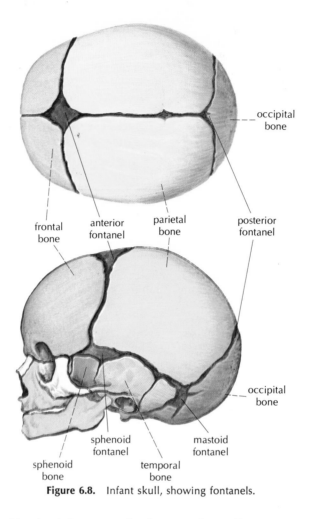

frontal
bone

anterior
fontanel

parietal
bone

posterior
fontanel

occipital
bone

occipital
bone

sphenoid
fontanel

mastoid
fontanel

sphenoid
bone

temporal
bone

Figure 6.8. Infant skull, showing fontanels.

lar) **cartilage**. Inside the joint space is the rather thick, colorless synovial fluid, which has been named after its resemblance to egg white.

The slightly movable and immovable joints form continuous structures in which either cartilage or fibrous connective tissue fills the gap between the bones. These soft tissue areas are larger in the child, become smaller in the adult, and may become completely filled by bone in later life. The joints between the vertebrae in the lower part of the spinal column and those between the three sections of each of the ossa coxae are examples of immovable joints that completely disappear rather early in life. The skull joints, or sutures, are held together by fibrous connective tissue aided by the dovetailing of the rather irregular sawtooth type of bone edges.

JOINT FUNCTION

The chief function of the freely movable joints is to allow for changes of position and so provide for motion. These movements are given names to describe the nature of the change in position of the body parts. Examples of some names of movements are:

1. **Flexion** (flek'shun), which is a bending motion, decreasing the angle between 2 parts.
2. **Extension**, which is a straightening or stretching: the reverse of flexion.
3. **Abduction**, which means motion *away from* the midline of the body. Raising the arm away from the body (as when reaching for something up on a high shelf) involves abduction of the upper extremity.
4. **Adduction**, which means motion *to-*

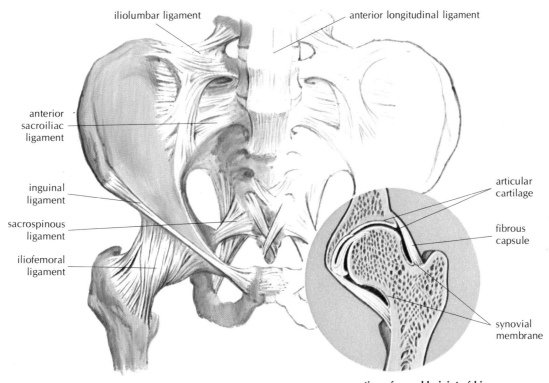

iliolumbar ligament

anterior longitudinal ligament

anterior
sacroiliac
ligament

inguinal
ligament

sacrospinous
ligament

iliofemoral
ligament

articular
cartilage

fibrous
capsule

synovial
membrane

cross section of movable joint of hip

Figure 6.9. Ligaments of hip and pelvis.

ward the midline. The scissors kick used in swimming is an example of adduction of the lower extremities.

5. **Rotation**, which means motion around a central axis, as in swinging the head in saying "No."

Muscles

CHARACTERISTICS OF SKELETAL MUSCLE

As noted in Chapter 2, there are three basic kinds of muscle tissue: skeletal, smooth and cardiac muscle. This chapter will be concerned only with skeletal muscle, which is attached to the bones, also known as voluntary muscle because it is normally under the control of the will (Figs. 6.10 and 6.11).

Skeletal muscles may be regarded as organs since they are made of a combination of muscle and connective tissue. When stimulated by nerve impulses, the long and threadlike muscle cells (fibers) become shorter and thicker, resulting in muscle contraction. Muscle fibers are arranged in bundles that are held together by connective tissue. Groups of these bundles are held together by additional connective tissue, and the entire muscle is encased in a tough connective tissue sheath called the **epimysium** (ep-i-mis'e-um) (Fig. 6.12).

Most muscles have two or more attachments to the skeleton. The method of attachment varies. In some instances the connective tissue within the muscle ties directly to the periosteum of the bone. In other cases the connective tissue sheath and partitions within the muscle all extend to form specialized structures that aid in attaching the muscle to bones. Such an extension may take the form of a cord, in which case it is called a **tendon**. In other cases a sheet called an **aponeurosis** (ap-o-nu-ro'sis), or else **fascia** (which has been discussed previously), may attach muscles to bones.

Whatever the nature of the muscle attachment, the principle remains the same: to furnish a means of harnessing the power

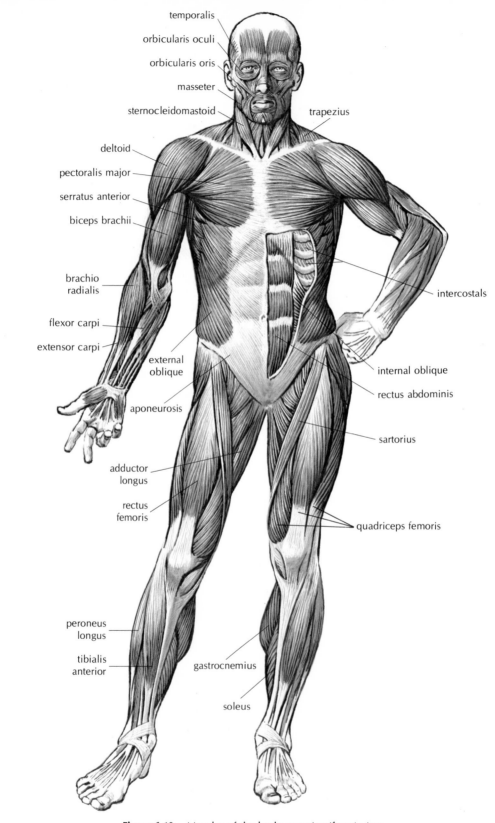

temporalis

orbicularis oculi

orbicularis oris

masseter

sternocleidomastoid

trapezius

deltoid

pectoralis major

serratus anterior

biceps brachii

brachio
radialis

flexor carpi

extensor carpi

external
oblique

aponeurosis

adductor
longus

rectus
femoris

peroneus
longus

tibialis
anterior

intercostals

internal oblique

rectus abdominis

sartorius

quadriceps femoris

gastrocnemius

soleus

Figure 6.10. Muscles of the body, anterior (front) view.

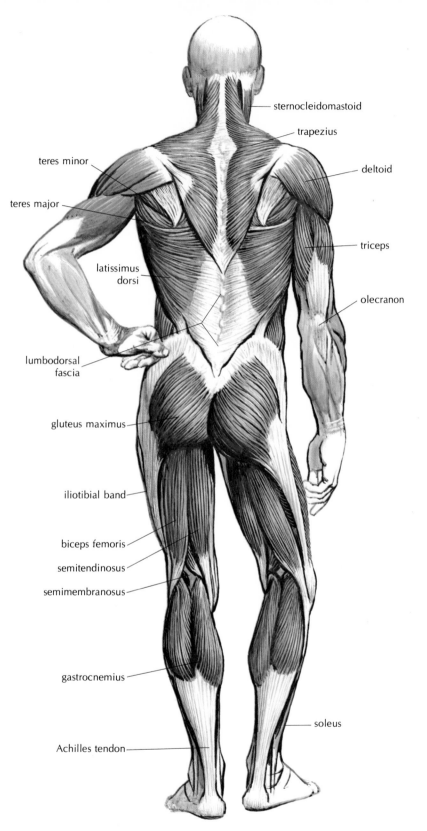

Figure 6.11. Muscles of the body, posterior (back) view.

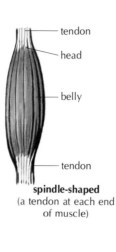

tendon

head

belly

tendon

spindle-shaped
(a tendon at each end
of muscle)

flat
(serrated margin,
broad tendon)

two-bellied
(each muscle is spindle
shaped, connected by
a short tendon)

two-headed
(biceps)
(each of its two heads
has its own muscle belly)

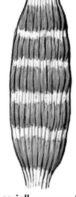

Figure 6.12. Types of muscles.

unipennate
(tendon along left margin)

serially arranged
(parallel rows of
intermediate tendons)

bipennate
(tendon in middle)

of the muscle contractions. A muscle has two (or more) attachments, one of which is more freely movable than the other. The less movable (more fixed) attachment is called the **origin**; the attachment to the part of the body which the muscle puts into action is called the **insertion**. In muscle contraction, the muscle is shortened, pulling on the two types of attachments and bringing them nearer to each other, thereby causing actual muscle movement, or muscle **action.**

Within the muscle sheath are found blood and lymphatic vessels as well as nerve fibers. Muscles receive an abundant blood supply because of the large amount of oxygen consumed by their cells, along with sugar, to generate power and heat.

Nerve fibers carry impulses to the muscles, each fiber supplying from a few up to more than a hundred individual muscle cells. The endings of the motor nerve fibers are called **myoneural** (mi-o-nu′ral) **junctions** or **motor end plates** (see Chapter 7).

Muscle tone refers to a partially contracted state of the muscles which is normal even though the muscles may not be in use at the time. The maintenance of this tension is due to the action of the nervous system, and the effect of it is to keep the muscles in a constant state of readiness for action. Muscles that are little used soon become flabby, weak and out of tone. If a person's muscle tone is poor, his general state of health is considered "below par."

SOME INDIVIDUAL SKELETAL MUSCLES
The human body contains more than 400 skeletal muscles, constituting between 35 and 40 percent of body weight. A large num-

ber of the skeletal muscles are arranged in pairs. A movement is initiated by one muscle or set of muscles called the **prime mover**. When an opposite movement is to be made, another muscle or set of muscles, known as the **antagonist**, takes over. In this way body movements are coordinated, and a large number of complicated movements are carried out without the necessity of planning in advance the means of performing them. At first, however, any new complicated movement must be learned. Think of a child learning to walk or to write, and consider the number of muscles which he uses unnecessarily or forgets to use when the situation calls for them.

Muscles of the head

The principal muscles of the head are those of facial expression and of chewing (mastication) (Fig. 6.13).

The muscles of facial expression include the vaguely circular ones around the eyes and the lips. They are called the **orbicularis** (or-bik-u-la′ris) **muscles** because of their shape (think of "orbit"). The muscle surrounding each eye is called the **orbicularis oculi** (ok′u-li), while the muscle of the lips is the **orbicularis oris**. These muscles, of course, are all provided with antagonists.

One of the largest muscles of expression forms the fleshy part of the cheek and is called the **buccinator** (buk′se-na-tor). It is used in whistling or blowing and is sometimes referred to as the trumpeter's muscle. You can readily think of other muscles of facial expression: for instance, the antagonists of the orbicularis oris which can produce a smile, a sneer or a grimace. There are a number of scalp muscles by means of which the eyebrows are lifted or else drawn together into a frown.

There are four pairs of muscles of chewing, all of which insert on the mandible and move it. The largest are the **temporal** (tem′po-ral) **muscles**, located above and near the ear, and the **masseter** (mas-se′ter) **muscles** at the angle of the jaw.

The tongue has two groups of muscles. The first group, called the **intrinsic muscles,** are located entirely within the tongue. The second group, the **extrinsic muscles,** origi-

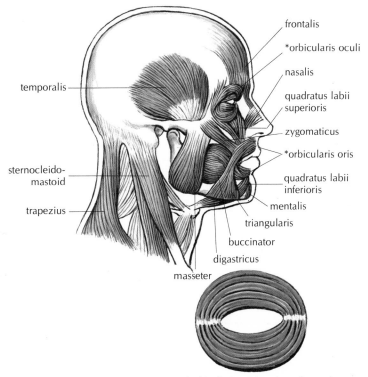

Figure 6.13. Muscles of the head.

*orbicularis or ring-shaped muscle

nate outside the tongue. It is because of these many muscles that the tongue has such remarkable flexibility and can perform so many different functions. Consider the intricate tongue motions involved in speaking, chewing and swallowing.

Muscles of the neck

The neck muscles tend to be ribbonlike and extend up and down or obliquely in several layers and in a complex manner. The two pairs you will hear of most frequently are the **trapezius** (trah-pe′ze-us) **muscles,** which extend into the upper back, and the **sternocleidomastoid** (ster-no-kli-do-mas′toid), sometimes referred to simply as the **sternomastoid.** The latter muscle extends along the side of the neck. It may be injured or for other reasons become shortened, resulting in a condition called wryneck (**torticollis**). Both of these muscles help to hold the head erect. The upper part of the trapezius also assists in moving the head to one side or the other.

Muscles of the upper extremities

The muscles that produce finger movements are the several **flexor digitorum** (dij-e-to′rum) and the **extensor digitorum muscles.** The thumb has a special muscle for flexion; and this appendage, incidentally, by virtue of its position and freedom of movement, has been one of the most marvelously useful endowments of man. Many movements of the wrist are produced by the **flexor carpi** and the **extensor carpi muscles.** The muscles of the fingers, the thumb and the wrist are located in the forearm with tendons extending into the wrist, hand and fingers, except for the smaller and less powerful muscles forming the few fleshy parts of the hands and fingers.

In the upper arm is the **biceps brachii** (bra′ke-i), the muscle almost invariably displayed by small boys as proof of their strength. On the opposite side of the upper arm is the **triceps brachii,** or boxer's muscle, which straightens the elbow when a blow is delivered.

The muscular cap of the shoulder is the **deltoid muscle,** an upside-down triangle that abducts the arm. The deltoid is often used as an injection site.

Muscles of the chest and the back

On either side of the chest, high up in the breast region, is the large **pectoralis** (pek-to-ra′lis) **major.** This muscle begins at the sternum, the upper ribs and the clavicle, and forms the anterior "wall" of the armpit. It ends at the humerus and serves to flex the arm across the chest.

Below the armpit, on the side of the chest, is the **serratus** (ser-ra′tus) **anterior.** It originates on the upper eight or nine ribs on the side and the front of the thorax and inserts in the scapula on the side toward the vertebrae. The serratus anterior muscle moves the scapula forward when, for example, one is pushing something.

In the spaces between the ribs are inserted the **intercostal muscles.** These are particularly important in respiration, serving to enlarge the thoracic cavity upon inspiration.

If we move around to the back of the body, the first important muscle to be encountered is the **trapezius,** part of which we have already seen in the neck. The trapezius is a large muscle which comes down from the neck and fans out like a pointed cape over the back of the shoulders. One purpose of the trapezius is to maintain the position of the shoulders. If you raise your shoulders and throw them back, the trapezius is the chief muscle used.

The other superficial muscle is the **latissimus** (lah-tis′i-mus) **dorsi,** originating in the lower half of the spine and covering the lower half of the thoracic region. This is the principal muscle used in bringing the arm down forcibly, as, for example, in swimming. The muscles which act on the vertebral column itself are thick vertical masses that lie under the relatively flat trapezius and latissimus dorsi.

Muscles of the abdomen

The abdominal muscles serve a number of purposes. One is to assist indirectly in the process of respiration, relaxing when the diaphragm contracts and vice versa. They are also compressors of internal organs, aiding in the acts of defecation, urination, childbirth and other processes. Another function of the abdominal muscles is to bend the trunk in various directions.

The lateral walls of the abdomen have three main muscles arranged in layers: the **external oblique** on the outside, the **internal oblique** in the middle, and the innermost layer called the **transversus abdominis**. The fibers of these three muscles all run in different directions. With these layers "glued" together, the total effect is like that of a piece of plywood; the result is a very strong abdominal wall. The front of the abdomen is closed in by the long, narrow **rectus abdominis**, which originates at the pubis and ends at the ribs. It is surrounded by connective tissue layers from the other three muscles.

Within the ventral body cavity, forming the partition between the abdominal cavity and the thoracic cavity, is the **diaphragm**, the dome-shaped muscle used in breathing.

The pelvic floor, or **perineum** (per-i-ne′um), has its own form of diaphragm, shaped somewhat like a shallow dish. One of the principal muscles of this pelvic diaphragm is the **levator ani** (le-va′tor a′ni), which acts on the rectum and thus aids in defecation.

Muscles of the lower extremities

The muscles in the lower extremities are among the largest and strongest in the body and are specialized for locomotion and balance. The **gluteus maximus** (gloo′te-us mak′si-mus), which forms much of the fleshy part of the hips and the buttocks, is relatively large in humans because of its function in standing in the erect position. This muscle extends the hip and is very important in walking and running. It is also used frequently as a site for injections, particularly if the quantity of the medication is large.

The **sartorius** is a long, narrow muscle that begins at the iliac spine, winds downward and inward across the entire thigh, and ends on the upper medial surface of the tibia. It is called the tailor's muscle because it is used in crossing the legs.

On the front of the thigh is a large four-part muscle called the **quadriceps femoris** (kwod′re-seps fem′or-is). This muscle extends the knee as in kicking a football. Charleyhorse is a spasm, or soreness and stiffness, which may involve any muscle, but this term is used most often in connection with the quadriceps femoris. Sometimes this muscle is used as an injection site.

A short double muscle located high up on the thigh is the powerful **iliopsoas** (il-e-o-so′as). This muscle extends from the lumbar vertebrae and the ilium to the top part of the femur. It is a flexor of the thigh and also helps to keep the trunk from falling backward when the body is standing erect.

On the inside (medial part) of the thigh are the **adductors**, which serve to press the thighs together. Anyone who rides horseback uses these muscles more than any others.

The posterior thigh muscles are called the **hamstring group**, and their tendons can be felt behind the knee on either side. They serve to bend the leg backward toward the thigh.

The **gastrocnemius** (gas-trok-ne′me-us) is the chief muscle of the calf of the leg. It has been called the toe dancer's muscle because it is necessary in order to stand on tiptoe. It ends near the heel in a prominent cord called the **Achilles tendon**, which then attaches to the calcaneus (heel bone). The Achilles tendon is the largest tendon in the body.

Another leg muscle which acts on the foot is the **tibialis** (tib-e-a′lis) **anterior**, located on the front of the leg. This muscle performs the opposite function of the gastrocnemius. Anyone who feels inclined to walk on his heels will use the tibialis anterior to raise the rest of the foot off the ground.

The ankle joint is not constructed exactly like the joint of the hand; it does not have quite the same freedom of movement. The ankle does have a small amount of "side play" (i.e., medial and lateral rotation). These movements are called: **inversion** (to bend the ankle so that the sole of the foot is facing the opposite foot) and **eversion** (bending the ankle in the opposite direction so that the sole of the foot is facing outward, away from the body). The principal muscle of inversion is the tibialis anterior, just discussed; the muscle of eversion is the **peroneus** (per-o-ne′us) **longus**, located on the outside of the leg (i.e., laterally).

The toes, like the fingers, are provided

Largest Muscles

NAME	LOCATION	FUNCTION
deltoid	cap of shoulder—upper arm	raises (abducts) arm
pectoralis major	front of chest to arm	adducts and flexes arm (antagonist of latissimus dorsi)
trapezius	back of neck and upper back	raises, lowers or rotates scapula (shoulder)
latissimus dorsi	middle and lower back (ends on upper arm)	adducts and extends arm (antagonist of pectoralis major)
biceps brachii	front of arm	flexes elbow (forearm)
triceps brachii	back of arm to elbow	extends forearm (antagonist of biceps brachii)
sacrospinalis	back, deep, vertical	extends head and spine
rectus abdominis	middle area of anterior abdominal wall	flexes spine and compresses abdominal cavity and its contents
gluteus maximus	hip superficially	extends thigh
iliopsoas	deep groin area	flexes thigh (antagonist of gluteus maximus)
adductor magnus	thigh medially	adducts thigh (as in horseback riding)
gluteus medius	hip laterally	abducts thigh (antagonist of adductor magnus)
quadriceps femoris	anterior thigh	strong extensor of knee
biceps femoris (one of 3 hamstrings)	posterior thigh	flexor of knee (antagonist of the quadriceps femoris)
gastrocnemius	calf of leg	extends foot (as in tiptoeing)
tibialis anterior	anterior and lateral shin region	flexes foot (dorsal flexion)

with a set of muscles likewise called the **flexor digitorum** and the **extensor digitorum muscles**. The former originate both at the tarsus and at the posterior surface of the tibia, and pass through the sole of the foot to the toes. The latter arise both from the tarsus and the anterior surface of the fibula, and pass over the top of the foot to the toes.

Applying Knowledge About the Musculoskeletal System

EXERCISE

The rate of muscle metabolism increases during exercise, resulting in a relative deficiency of oxygen and muscle nutrients. In turn, this deficiency causes **vasodilation** (vas-o-di-la′shun)—an increase in caliber of blood vessels—thereby allowing blood to flow easily back to the heart. The temporary increased load on the heart acts to strengthen heart muscle and to improve the circulation within the heart muscle. Regular exercise also improves respiratory efficiency; circulation in the capillaries surrounding the alveoli, or air sacs, is increased, and this brings about enhanced gas exchange and deeper breathing.

LEVERS AND BODY MECHANICS

Proper body mechanics helps to conserve energy and to ensure freedom from strain and fatigue; conversely, such ailments as lower back pain—a very common complaint—can be traced to poor body mechanics. Body mechanics has special significance to health workers, who are frequently called upon to move patients, handle cumbersome equipment and so on. Maintaining the body segments in correct relation to one another has a direct effect on the working capacity of the vital organs that are supported by the skeletal structure.

If you have had a course in physics, recall your study of levers. A lever is simply a machine consisting of a rigid bar that moves about a fixed point, the fulcrum. There are three classes of levers, which differ only in the location of the fulcrum, the effort (force) and the resistance (weight). In a first class lever, the fulcrum is located between the resistance and the effort; scissors, which you probably use every day, are an example of this class. The second class lever has the resistance located between the fulcrum and the effort; turning a mattress, and lifting one end of a bed, are illustrations of this class. In the third class lever the effort is between the resistance and the fulcrum; your arm supporting an object held in your hand is an example of this class of lever. The musculoskeletal system can be considered a system of levers.

Bursae and Foot Arches

A **bursa** (bur'sah) is a padlike sac containing a small amount of synovial fluid, that is located in an area where friction may develop. Several are distributed about the knee joint and others are found in the region of the shoulder, the elbow and the hip. **Bursae** (bur'se) are also found between tendons and between muscles, serving as slippery cushions to prevent friction between moving parts.

The three arches of the foot are supported by muscles, tendons, and ligaments. They provide strong arched levers for the feet, and so are important in walking, running and jumping. The arches of the foot are:

1. The **medial longitudinal arch** (the most definite arch) which extends the length of the foot on the inside (medial) edge.
2. The **lateral longitudinal arch** which extends along the outside (lateral) edge of the foot.
3. The transverse or **metatarsal** arch which extends crosswise under the ball of the foot, near the toes.

Summary

1. **Bones.**
   A. Specific bones.
      (1) Head.
         (a) Cranial: frontal, 2 parietal, 2 temporal, ethmoid, sphenoid, occipital.
         (b) Facial: mandible, 2 maxillae, 2 zygomatic, 2 lacrimal, vomer, 2 palatine, 2 nasal conchae, 2 nasal bones, ossicles (of ear), hyoid.
      (2) Trunk.
         (a) Vertebral column. Regions: cervical, thoracic, lumbar, sacral, coccygeal. Curves: primary and secondary.
         (b) Ribs: true ribs (first 7), false costae (5 plus 2 floating ribs).
      (3) Extremities.
         (a) Upper: pectoral girdle (clavicle, scapula), humerus, radius, ulna, 8 carpal bones, 5 metacarpal, phalanges.
         (b) Lower: pelvic girdle (2 ossa coxae), femur, patella (largest sesamoid), tibia, fibula, 7 tarsal, 5 metatarsal (calcaneus, etc.), phalanges.
   B. Landmarks of bones.
      (1) Prominences: olecranon, iliac spine, ischial spine, trochanters.
      (2) Pelvic divisions: ilium, ischium, pubis.
      (3) Anterior fontanel (and others) in infants.
      (4) Foramina: foramen magnum, obturator foramina.
      (5) Fossae and grooves.
2. **Joints.**
   A. Types: immovable, slightly movable, freely movable.
   B. Structure: held together with ligaments; surfaces covered with articular cartilage, lubricated with synovial fluid.
   C. Movements: flexion, extension, abduction, adduction, rotation.
3. **Muscles.**
   A. Characteristics: bundles of fibers covered with epimysium; attached by tendons or aponeuroses (or fasciae); fixed attachment (origin), movable attachment (insertion); nerve fibers end at motor end plates; muscle tone maintained by nervous system.
   B. Individual muscles.
      (1) Head.
         (a) Face: orbicularis oculi, orbicularis oris, buccinator, scalp muscles (some).
         (b) Mastication: temporal, masseter.
         (c) Tongue: intrinsic, extrinsic.
      (2) Neck: trapezius, sternocleidomastoid.
      (3) Upper extremities: flexor digitorum and extensor digitorum muscles; flexor carpi and extensor carpi muscles; biceps brachii, triceps brachii, deltoid.

(4) Chest and back: pectoralis major, serratus anterior, intercostals, trapezius, latissimus dorsi.

(5) Abdomen: external oblique, internal oblique, transversus abdominis, rectus abdominis, diaphragm, pelvic floor muscles (including levator ani).

(6) Lower extremities: gluteus maximus, sartorius, quadriceps femoris, iliopsoas, adductors, hamstring group, gastrocnemius, tibialis anterior, peroneus longus, flexor digitorum and extensor digitorum muscles.

C. Table of largest muscles.

D. Practical applications: value of exercise; importance of correct posture; bones and muscles as levers.

E. Bursae and foot arches.

(1) Bursae: prevent friction.

(2) Arches: medial longitudinal most important.

Questions and Problems

1. Give a general description of a bone with respect to its tissues, membranes, vessels, etc.
2. Name 5 general functions of bones.
3. What are the main cranial and facial bones?
4. What are the main divisions of the vertebral column? The ribs?
5. Name the bones of the upper and lower extremities.
6. Name: 4 bone prominences; 3 divisions of the pelvic girdle; the largest fontanel; 2 prominent foramina.
7. What are the 2 kinds of joints?
8. Describe joint structure and describe 5 kinds of joint movements.
9. Describe 3 different joint diseases.
10. Give a general description of skeletal muscle with respect to its structure, function, attachments and nerve connections.
11. Name the main muscles of the head and define each. What are the 2 principal neck muscles?
12. Name and define the principal muscles of the upper extremities; the chest and the back.
13. Name the main muscles of the abdominal wall and give the reason for their especially light and strong construction.
14. Name and locate the main lower extremity muscles.
15. Name 3 pairs of muscle antagonists.
16. What are some valuable effects of exercise?
17. What is one way of checking one's posture?
18. How does sitting in a chair affect posture?
19. What are levers and how do they work?
20. The forceps is an example of which class of levers?
21. Define a bursa and indicate some locations of bursae.
22. Name the 3 arches of the foot and describe their function.

The Brain, the Spinal Cord and the Nerves

Once a number of the body systems have been surveyed, it should become fairly obvious that not one of these systems is capable of functioning alone. All are interdependent, and all must work together as one unit in order that the normal conditions within the body may prevail. The agency that insures the coordination of the organs and organ systems is the nervous system. Conditions both within and outside the body are constantly changing; one purpose of the nervous system is to respond to these internal and external changes (known as **stimuli**) and so cause the body to adapt itself to new conditions.

The nervous system has been very aptly compared with a telephone exchange in which the brain and the spinal cord act as centers and the nerve trunks serve as cables and wires for carrying messages to and from various parts of these centers.

The Nervous System as a Whole

The parts of the nervous system may be grouped according to how they are made (structure) or else on the basis of what they do (function).

STRUCTURAL (ANATOMIC) CLASSIFICATION
The structure of the nervous system serves as the basis for the more commonly used grouping of the parts of the nervous system, as follows:

1. The **central** nervous system, which includes the brain and the spinal cord.

2. The **peripheral** nervous system, which is made up of **cranial** and **spinal** nerves. Cranial nerves are those which carry impulses to and from the brain. Spinal nerves are those which carry messages to and from the spinal cord.

From the standpoint of structure, the central and peripheral nervous systems together include most of the nerve tissue in the body. However, certain peripheral nerves have a special function, and for this reason they are grouped together under the designation **autonomic** (aw-to-nom'ik) nervous system. The reason for this separate classification is that the autonomic nervous system has to do largely with activities which go on more or less automatically. This system carries impulses from the central nervous system to the glands, the involuntary muscles found in the walls of tubes and hollow organs and the heart. The autonomic nervous system is subdivided into the **sympathetic** and **parasympathetic** nervous systems, both of which will be explained later in this chapter.

Some of the nerves that carry autonomic nervous system impulses are cranial, and others are spinal.

On Nerves in General

The basic nerve cell is called a **neuron**. Neurons are composed of a **cell body**, containing the nucleus, with the addition of threadlike projections of the cytoplasm known as **nerve fibers**. The nerve fibers are of two kinds: **dendrites** (den'drites), which conduct impulses *to* the cell body; and **axons**, which conduct impulses *away from* the cell body. The dendrites of sensory neurons are very different from those of other neurons. They are usually single and they may be very long (as much as 3 feet) or they may be short; but in any case, they do not have the treelike appearance so typical of other dendrites. Each sensory nerve fiber (dendrite) has a special structure called the **receptor**, or **end organ**, where the stimulus is received and the sensory impulse begins. Sensations such as pain, touch, hearing and seeing which involve these sensory neurons will be discussed in Chapter 8 (see Figs. 7.1 and 7.2).

Each neuron is a separate unit, and there is no anatomic unity between neurons. It would be logical to ask how it is possible for neurons to be in contact; in other words, how the axon of one neuron can be in functional contact with the dendrite of another neuron. This is accomplished by the **synapse** (si'naps), from a Greek word meaning "to clasp." Synapses, then, are points of junction for transmission of nerve impulses.

A **nerve** is a bundle of nerve fibers, located *outside* the central nervous system, which conducts impulses from one place to another. A nerve can be compared with a telephone cable made up of many wires. In the case of nerves the "wires," or nerve fibers, are bound together with connective tissue.

Nerve fibers that are connected with receptors (for receiving stimuli) conduct impulses *to* the brain and cord, and when grouped together form **afferent** nerves. Those fibers that carry impulses *from* the centers out to the muscles and glands form **efferent** nerves. Some nerves contain a mixture of afferent and efferent nerve fibers and are often referred to as **mixed nerves**.

The Central Nervous System

THE BRAIN
Main parts of the brain

The largest part of the brain is the **cerebrum** (ser'e-brum), which is divided into the two **cerebral** (ser'e-bral) **hemispheres**, a right and a left one. The **brainstem** includes the deeper parts that comprise the interbrain (thalamus, etc.) that cannot be seen unless the brain is sectioned, and a series of smaller parts that extend downward. Starting at the upper part of this series, we may see a small part of the **midbrain**. Below it and plainly visible from the under view of the brain are the **pons** (ponz) and the **medulla oblongata** (me-dul'lah ob'long-ga'tah), which connects with the spinal cord through a large opening in the base of the skull. The pons connects the midbrain and the medulla. Next in size to the cerebral hemispheres is the **cerebellum** (ser-e-bel'um), a word meaning "little brain." It is located immediately below the back part of the cerebral hemispheres and is con-

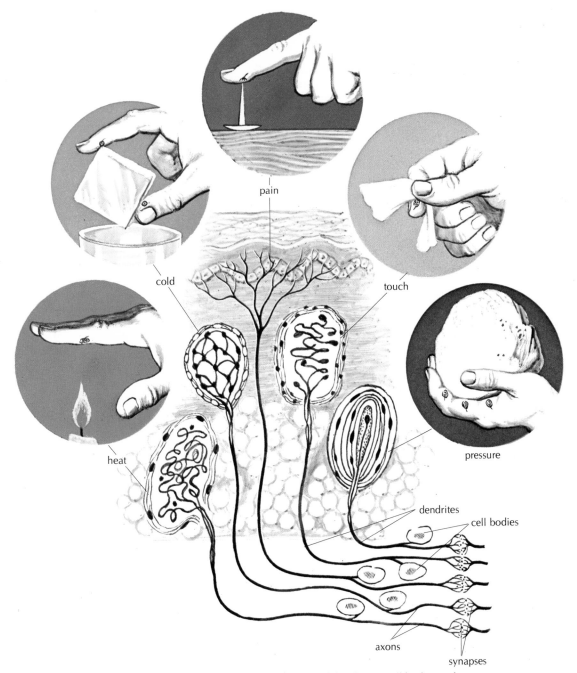

Figure 7.1. Diagram showing the superficial receptors and the deeper cell bodies and synapses to suggest the continuity of sensory pathways into the central nervous system.

nected with the other parts of the brain only by means of the bridgelike pons.

Structure of the cerebral hemispheres

The outer nerve tissue of the cerebral hemispheres is gray matter and is called the **cerebral cortex**. This gray cortex is arranged in folds forming elevated portions known as **convolutions** (kon-vo-lu'shuns), separated by depressions or grooves called **fissures**, or **sulci** (sul'si). Internally the cerebral hemispheres are made largely of white matter

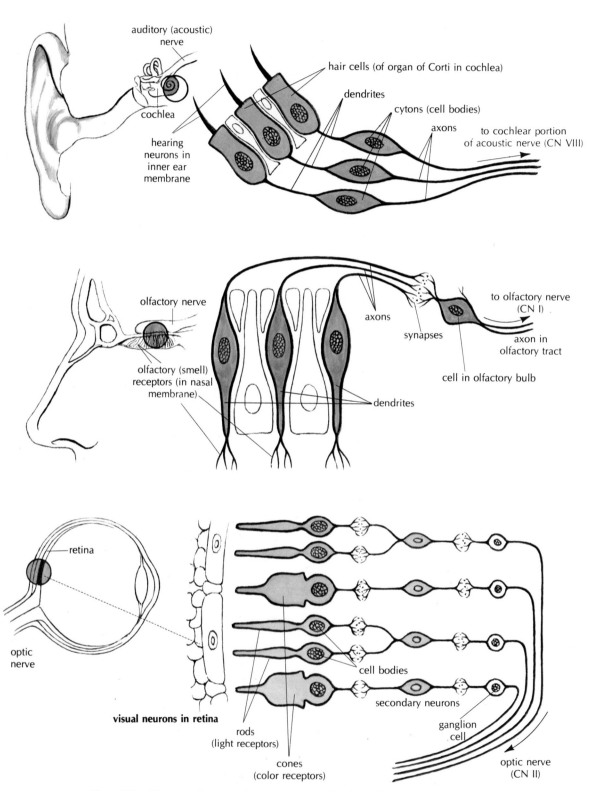

Figure 7.2. Diagram of neurons for receiving impulses from the special sense organs.

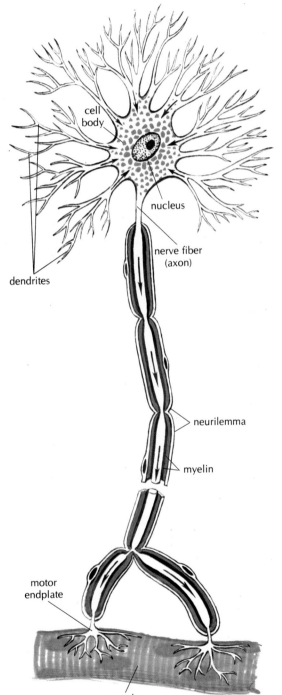

Figure 7.3. Diagram of a motor neuron.

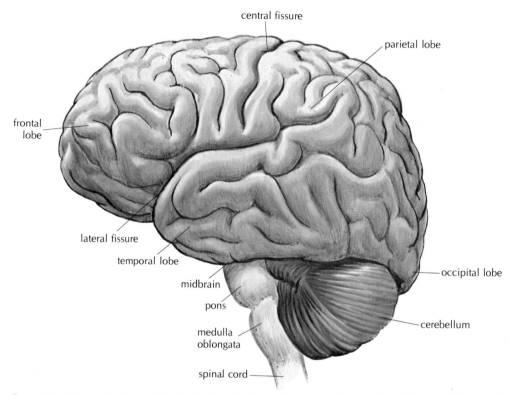

central fissure

parietal lobe

frontal lobe

lateral fissure

temporal lobe

midbrain

pons

medulla oblongata

spinal cord

occipital lobe

cerebellum

Figure 7.4. External surface of the brain showing the main parts and some of the lobes and fissures of the cerebrum.

and a few islands of gray matter. Inside the hemispheres are two spaces extending in a somewhat irregular fashion. These are the **lateral ventricles**, which are filled with a watery fluid common to both the brain and the spinal cord called **cerebrospinal** (ser-e-bro-spi′nal) **fluid**, to be discussed later.

Although there are many fissures (sulci), a few are especially important landmarks. These include the:

1. **Longitudinal fissure**, which is a deep groove that separates the upper parts of the cerebral hemispheres from each other.
2. **Central fissure**, which extends from the top of the brain near the center downward along the side at right angles to the longitudinal fissure.
3. **Lateral fissure**, which curves somewhat along the side of the brain and separates the temporal lobe from the rest of the cerebral hemisphere (see Fig. 7.4).

Let us examine the cerebral cortex, the layer of gray matter which forms the sur-

face of each cerebral hemisphere. It is within the cerebral cortex that all impulses are received and analyzed. These form the basis of knowledge; the brain "stores" knowledge, much of which can be produced on demand by means of the phenomenon which we call memory. It is in the cerebral cortex that all thought takes place, all association, judgment and discrimination. It is from the cerebral cortex, too, that the orders originating from conscious deliberation emanate; that is, the voluntary movements are controlled here (Fig. 7.5).

Division and functions of the cerebral cortex

The cerebral cortex of each hemisphere is divided into four **lobes**, areas named from the overlying cranial bones. It has been found that each area controls a certain category of functions. The four lobes, with some of their characteristic functions, follow:

1. The **frontal lobe**, which is relatively much larger in the human being than

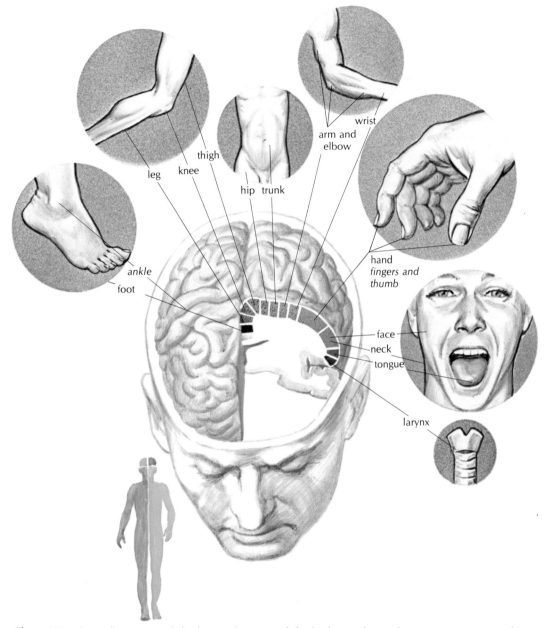

thigh

leg   knee

hip trunk

wrist

arm and
elbow

ankle

foot

hand
*fingers and
thumb*

face
neck
tongue

larynx

Figure 7.5. Controlling areas of the brain. The parts of the body are shown drawn in proportion to the area of control.

in any other organism. This contains the **motor cortex** which controls the voluntary muscles. The left side of the brain governs the right side of the body and the right side of the brain governs the left side of the body. The upper portion of the center controls the lower parts of the body. The frontal lobe also contains two

areas used in speech (the speech centers will be discussed later).

2. The **parietal lobe**, which occupies the upper part of each hemisphere, just behind the central fissure. This contains the **sensory area**, in which the general senses such as pain, touch and temperature are interpreted. Also, such interpretations as the determina-

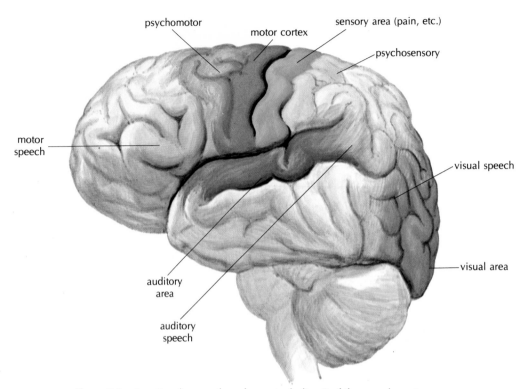

Figure 7.6. Functional areas of cerebrum, including 3 of the speech centers.

tion of distances, sizes and shapes take place here.

3. The **temporal lobe**, which is lateral (at the side) and folds under the hemispheres on each side. This contains the **auditory center** for interpreting impulses from the ear.

4. The **occipital lobe**, which is the most posterior, and extends over the cerebellum. This contains the **visual area** for interpreting messages from the retina of the eye.

Speech centers

The speech centers are among the most interesting groups of areas in the cerebral hemispheres (Fig. 7.6). Their development and use are closely connected with the processes of learning. These areas are called the:

1. **Auditory speech center**, located in the temporal lobe near the auditory center. While the auditory center enables a person to interpret sounds, it does not have anything to do with the understanding of words. Such an understanding of language requires the development of the auditory speech center. Often this is the first speech center to develop in the child. Babies often seem to understand what is being said long before they do any talking themselves. It is usually several years before children learn to read or write words. In many parts of the world people never learn to read or write their language.

2. **Visual speech center**, which is somewhat above and in front of the visual center. In this area the ability to read with understanding is developed. You may see the writing in the Japanese language, for example, but this would involve only the visual center in the occipital lobe unless you could read the words.

3. **Motor speech center**, located just in front of the lowest part of the motor cortex in the frontal lobe. Since the lower part of the motor cortex controls the muscles of the head and the neck, it seems logical to think of the

speech area as an extension forward to make possible the control of the muscles of speech in the tongue, the soft palate and the larynx.

4. **Written speech center,** located above the motor speech center, and in front of the cortical area that controls the muscles of the arm and the hand. Again this center is an extension forward from the motor cortex. The ability to write words usually is one of the last phases in the development of learning words and their meaning, although occasionally a person may write words more readily than he can vocalize them.

Other parts of the cerebral hemispheres

Beneath the gray matter of the cerebral cortex is the white matter, consisting of nerve fibers which connect the cortical areas with each other and with other parts of the nervous system. Among the most important of these large collections of nerve fibers is the **internal capsule,** a crowded strip of white matter where any injury is apt to cause extensive damage. At the base of each hemisphere are the nerve cell groups called **basal ganglia** (gang'gle-ah), which regulate the body movements originating in the cerebral cortex. On the underside of each cerebral hemisphere is the **olfactory** (ol-fak'to-re) **area,** concerned with the sense of smell, which is stimulated by the impulses arising in the nerve receptors of the nose.

The interbrain

The interbrain, or **diencephalon** (di-en-sef'ah-lon), can be seen only by cutting into the central section of the brain. It includes the **thalamus** and the **hypothalamus.** The two masses of gray matter that form the thalamus are relay centers and act to monitor sensory stimuli, suppressing some and magnifying others. The hypothalamus is located in the midline area below the thalamus and contains cells that control body temperature, water balance, sleep, appetite and some of our emotions, such as fear and pleasure. Both divisions of the autonomic nervous system are under the control of the hypothalamus. Thus it influences the heart's beating, the contractions of the walls of the bladder and other vital body functions.

The midbrain

The midbrain is located just below the center of the cerebrum. It forms one of the forward parts of the brainstem. Four rounded masses of gray matter that are hidden by the cerebral hemispheres form the upper part of the midbrain. These four bodies, the **corpora quadrigemina** (kor'po-rah kwod-ri-jem'i-nah), act as relay centers for certain eye and ear reflexes. The ventral white matter of the midbrain conducts impulses between the higher centers of the cerebrum and the lower centers in the pons, cerebellum, medulla and spinal cord.

The cerebellum

The cerebellum is made up of three parts: the middle portion, called the **vermis** (meaning "wormlike"), and two lateral hemispheres at the sides. As in the case of the cerebral hemispheres, the cerebellum has an outer area of gray matter and an inner portion that is largely white matter. The functions of the cerebellum are:

1. To aid in the coordination of voluntary muscles so that they will function smoothly and in an orderly fashion. Disease of the cerebellum causes muscular jerkiness and tremors.
2. To help maintain balance in standing, walking and sitting, as well as during more strenuous activities. Messages from the internal ear and from the tendon and muscle sensory end organs aid the cerebellum.
3. To aid in maintaining muscle tone so that all muscle fibers are slightly tensed and ready to produce necessary changes in position as quickly as may be necessary.

The pons

The pons is white in color because it is composed largely of myelinated nerve fibers. These fibers in the pons carry messages from one side of the cerebellum to the other, from the cerebellum to the higher centers in the cerebrum and midbrain, and from the cerebellum to the lower centers, including the medulla and the spinal cord.

Not only is the pons an important connecting link between the cerebellum and the rest of the nervous system, but it also contains connections with four pairs of cranial nerves. Further it contains nerve fibers that carry impulses to and from the centers located above and below it. Certain reflex (involuntary) actions are controlled in the pons; namely, some occurring in respiration.

The medulla oblongata

The medulla of the brain is located between the pons and the spinal cord. It appears white externally because, like the pons, it contains many covered (myelinated) nerve fibers. Internally it contains collections of cell bodies (gray matter), which are called **centers** or **nuclei**. Among these are the very important vital centers including:

1. The **respiratory center**, which controls the muscles of respiration in response to chemical and other stimuli.
2. The **cardiac center**, which tends to slow the heart rate so that it will not beat too rapidly to be effective.
3. The **vasomotor** (vas-o-mo'tor) **center**, which affects the muscles in the blood vessel walls and hence helps to determine blood pressure.

The last four pairs of cranial nerves are connected with the medulla. The nerve fibers that carry messages up through the spinal cord to the brain continue through the medulla also, as do similar descending or **motor fibers**. These groups of nerve fibers form tracts (bundles) and are grouped together according to function. The motor fibers from the motor cortex of the cerebral hemispheres extend down through the medulla, and most of them cross from one side to the other (decussate) while going through this part of the brain. It is in the medulla that the shifting of nerve fibers occurs which causes the right cerebral hemisphere to control muscles in the left side of the body, and the upper portion of the cortex to control of muscles in the lower portions of the person. The medulla is an important reflex center, and it is here that certain neurons end and impulses are relayed to other neurons.

Ventricles of the brain

Within the brain are four fluid-filled spaces called the ventricles. These extend into the various parts of the brain in a somewhat irregular fashion. We have already mentioned the largest, the lateral ventricles in the two cerebral hemispheres. Their extensions into the lobes of the cerebrum are called **horns** (see Fig. 7.7). These paired ventricles communicate with a midline space, the third ventricle, by means of the openings called **foramina** (fo-ram'i-nah). The third ventricle is bounded on each side by the two parts of the thalamus, while the floor is occupied by the hypothalamus. Continuing down from the third ventricle a small canal, called the **cerebral aqueduct,** extends through the midbrain into the fourth ventricle. The latter is continuous with the near microscopic neural, or central, canal of the spinal cord. Do not confuse this tiny canal inside the cord with the much larger vertebral, or spinal, canal that is a part of the dorsal cavity enclosing the entire cord, together with its membranes and surrounding fluid. In the roof of the fourth ventricle are three openings that allow the escape of fluid to the area that surrounds the brain and spinal cord. This will be discussed later.

After removal of some of the fluid, air or other substances may be injected, and x-rays called **encephalograms** or **ventriculograms** are taken. Tumors or other brain disorders may sometimes be located by this means.

Brain waves

The interactions of the billions of nerve cells in the brain give rise to measurable electric currents. These may be recorded by an instrument called the **electroencephalograph** (e-lek-tro-en-sef'ah-lo-graf). The recorded tracings or brain waves produce an **electroencephalogram**, not to be confused with the encephalogram mentioned in the last paragraph.

THE SPINAL CORD

Location of the spinal cord

In the embryo the spinal cord occupies the entire spinal canal and so extends down into the tail portion of the vertebral col-

86

umn. However, the column of bone grows much more rapidly than the nerve tissue of the cord, so that the end of the cord soon fails to extend into the lower part of the spinal canal. This disparity in growth increases so that in the adult the cord ends in the region just below the area to which the last rib attaches (between the first and second lumbar vertebrae).

Structure of the spinal cord

Examination of the spinal cord reveals that it has a small irregularly shaped internal section made of gray matter (nerve cell bodies), and a larger area surrounding this gray part that is made of white matter (nerve fibers). A cross section of the cord shows that the gray matter is so arranged that a column of cells extends up and down dorsally, one on each side; another column is found in the ventral region; while a third less conspicuous part is situated on each side. These three pairs of columns of gray matter give this cross section an H-shaped appearance (see Fig. 7.8). The white matter can be seen to be made of thousands of nerve fibers arranged in three areas external to the gray matter on each side.

Functions of the spinal cord

The functions of the cord may be divided into three aspects:

1. Reflex activities, which involve the transfer and integration of messages

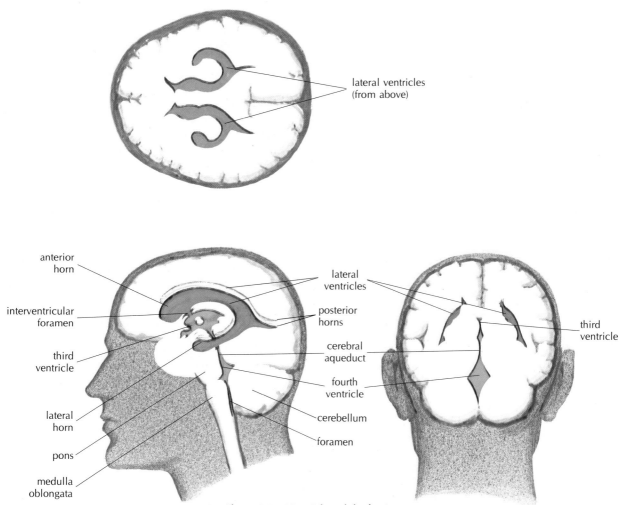

Figure 7.7. Ventricles of the brain.

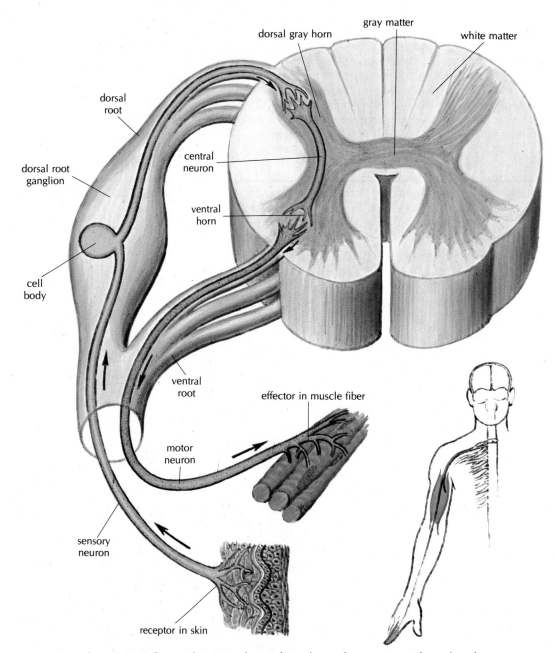

Figure 7.8. Reflex arc showing pathway of impulses and cross section of spinal cord.

that enter the cord, so that a sensory (afferent) impulse entering the center will become a motor (efferent) message leaving the cord.

2. A pathway for conducting sensory impulses from afferent nerves upward through ascending tracts to the brain.

3. A pathway for conducting motor (ef-

ferent) impulses from the brain down through descending tracts to the nerves that will supply muscles or glands.

The reflex pathway through the spinal cord usually involves three or more nerve cells together with their fibers, including:

1. The **sensory neuron**, which has its

beginning in a receptor and its nerve fiber in a nerve that leads to the cord (see Figs. 7.1 and 7.8).

2. One or more **central neurons**, which are entirely within the cord.

3. The **motor neuron**, which receives the impulse from a central neuron and then carries it via its long axon through a nerve to a muscle or a gland (see Figs. 7.3 and 7.8).

The knee jerk is an example of a spinal reflex. The pathway for the impulses that make this reflex possible includes a sensory neuron which has its receptor in the tendon just below the knee, its sensory nerve fiber in the nerves that extend to the spinal cord, central neurons inside the lower part of the cord and motor neurons that send processes through nerves from the cord to the effectors in the thigh's kicking muscle.

COVERINGS OF THE BRAIN AND THE SPINAL CORD

The **meninges** (me-nin′jez) are three layers of connective tissue that surround the brain and the spinal cord to form a complete enclosure. The outermost of these membranes is called the **dura mater** (du′rah ma′ter). It is the thickest and toughest of these meninges. Inside the skull, the dura mater splits in certain places to provide channels for the blood coming from the brain tissue. The second layer around the brain and the spinal cord is the **arachnoid** (ah-rak′noid) **membrane** (so-called because it resembles the webs produced by spiders, which belong to a group of animals called the Arachnida). The arachnoid membrane is loosely attached to the deepest of the meninges by weblike fibers allowing a space for fluid located between the arachnoid and the innermost membrane. The third layer around the brain, the **pia mater** (pi′ah ma′ter), is attached to the nerve tissue of the brain and spinal cord and dips into all the depressions, unlike the other two meninges. It is made of a delicate connective tissue in which there are many blood vessels. The blood supply to the brain is carried, to a large extent, by the pia mater (see Fig. 7.9).

CEREBROSPINAL FLUID

Fluids may serve to cushion shocks that would otherwise injure delicate organs. This is true of the normal "water on the brain" called cerebrospinal fluid. This fluid is formed inside the ventricles of the brain, mostly by special structures called the **choroid plexuses**. As in the case of other tissue fluids, some may be formed by filtration from the capillaries and may bring nutrients to the cells as well as remove waste products from the cells. This fluid normally flows freely from ventricle to ventricle and finally out into the **subarachnoid space**, which completely encloses the brain and spinal cord. Much of the fluid is returned to the blood in the venous sinuses through projections called the **arachnoid villi** (or granulations) (see Fig. 7.10).

The Peripheral Nervous System

CRANIAL NERVES
Location of the cranial nerves

There are 12 pairs of cranial nerves in all (henceforth, when a cranial nerve is referred to, a pair is really meant). The first four cranial nerves are located near the front of the brain and are attached to the undersurface. The next four nerves are related to the pons, and the last four are attached to the medulla. All but two of these nerves supply nerve fibers to structures in the head. One nerve extends to muscles in the neck, while another (the longest) cranial nerve sends branches into the thoracic and the abdominal organs (Fig. 7.11).

General functions of the cranial nerves

From a functional point of view we may think of the kinds of messages which the cranial nerves handle as belonging to one of four categories:

1. Special sense impulses such as smell, visual and hearing messages.
2. General sense impulses such as pain, touch, temperature, deep muscle sense, and pressure and vibration senses.
3. Voluntary muscle control or somatic motor impulses.
4. Involuntary control or visceral effec-

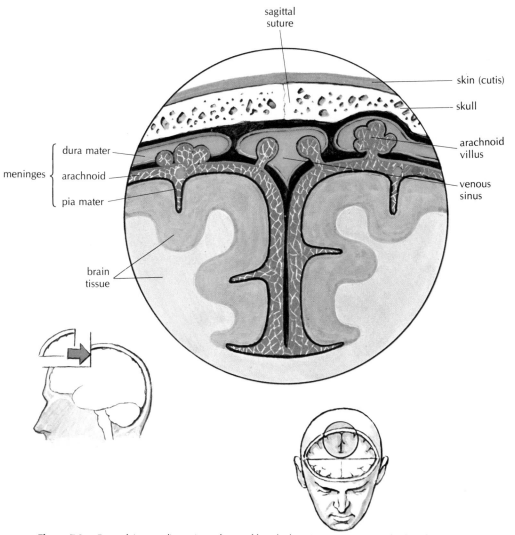

Figure 7.9. Frontal (coronal) section of top of head, showing meninges and related parts.

tor messages to glands and involuntary muscles.

Names and specific functions of the cranial nerves

A listing of the 12 cranial nerves follows:

I. The **olfactory nerve**, which carries smell impulses to the brain.

II. The **optic nerve**, which is the nerve of vision.

III. The **oculomotor nerve** which is concerned with the contraction of most of the eye muscles.

IV. The **trochlear** (trok'le-ar) **nerve**, which supplies one eyeball muscle on each side.

V. The **trigeminal** (tri-jem'i-nal) **nerve**, which is the great sensory nerve of the face and head. It has three branches that carry general sense impulses. The third branch is joined by motor fibers to the muscles of chewing (mastication).

VI. The **abducens** (ab-du'senz) **nerve** which is another nerve sending controlling impulses to an eyeball muscle.

VII. The **facial nerve,** which is largely

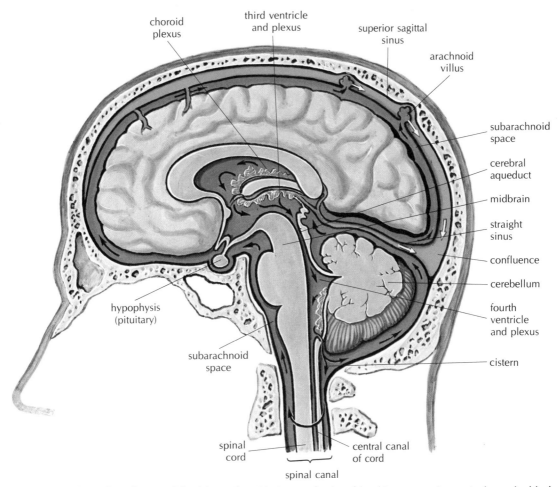

Figure 7.10. Flow of cerebrospinal fluid from choroid plexuses back to blood in venous sinuses is shown by black arrows; blood flow is shown by white arrows.

motor. The muscles of facial expression are all supplied by branches from the facial nerve. This nerve also includes sensory fibers for taste (forward two thirds of the tongue) and contains secretory fibers to the smaller salivary glands (the submaxillary and sublingual), and to the lacrimal gland.

VIII. The **acoustic nerve**, which contains special sense fibers for hearing as well as those for balance from the semicircular canals of the internal ear.

IX. The **glossopharyngeal** (glos-o-fah-rin′ge-al) **nerve** which contains general sense fibers from the back of the tongue and the pharynx (throat). This nerve contains all the four kinds of fibers. These include fibers for taste from the back of the tongue (posterior one third), those that supply the largest salivary gland (parotid), and motor nerve fibers to control the swallowing muscles (in the pharynx).

X. The **vagus** (va′gus) **nerve**, which is the longest cranial nerve. It supplies most of the organs in the thoracic and abdominal cavities. This nerve also contains secretory fibers to glands that produce digestive juices and other secretions.

XI. The **accessory nerve** (formerly called the spinal accessory nerve), which is made up of motor nerve

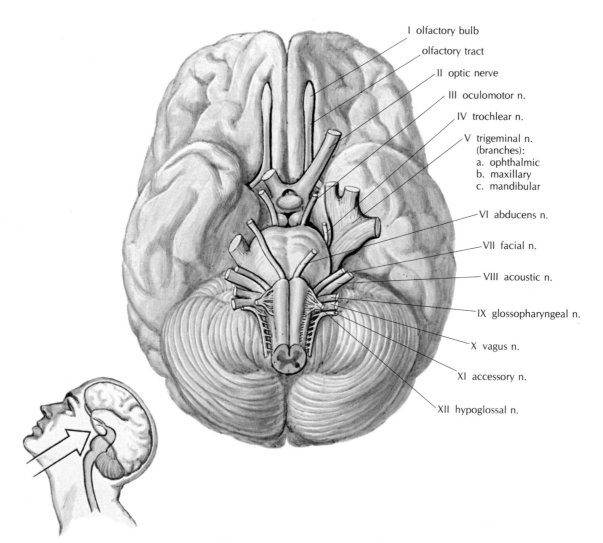

I olfactory bulb
olfactory tract
II optic nerve
III oculomotor n.
IV trochlear n.
V trigeminal n.
(branches):
a. ophthalmic
b. maxillary
c. mandibular
VI abducens n.
VII facial n.
VIII acoustic n.
IX glossopharyngeal n.
X vagus n.
XI accessory n.
XII hypoglossal n.

Figure 7.11. Base of brain, showing cranial nerves.

fibers controlling two muscles of the neck (trapezius and sterno-cleidomastoid).

XII. The **hypoglossal nerve**, the last of the 12 cranial nerves, which carries impulses controlling the muscles of the tongue.

SPINAL NERVES

*Location and structure
of the spinal nerves*

There are 31 pairs of spinal nerves, and each nerve is attached to the spinal cord by two roots: the **dorsal root** and the **ventral root**. To each dorsal root are attached small masses of nerve cell bodies (gray matter) called **dorsal root ganglia**, ganglia being defined as collections of nerve cell bodies usually located *outside* the central nervous system. To these ganglia lead nerve fibers from the sensory receptors of various areas of the body. A sensory receptor is a nerve ending which responds to stimuli. There are two categories of receptors: those which are located generally over the body and respond to general sensations such as pain, touch and temperature changes. Then there are the receptors of special sensations—taste, vision and so forth—from which impulses are carried by cranial nerves. These organs of special sense will be taken up in the next chapter.

The ventral roots are a combination of motor (efferent) nerve fibers supplying voluntary muscles, involuntary muscles and glands. The cell bodies for the voluntary fibers are located in the ventral part of the cord gray matter (anterior or ventral gray horn) while the cell bodies for the involuntary fibers are to be found in the lateral gray horns.

Branches of the spinal nerves

Each spinal nerve continues only a very short distance away from the spinal cord and then branches into small posterior divisions and rather large anterior divisions. The larger anterior branches interlace to form networks called **plexuses** (plek'sus-es) which then distribute branches to the body parts. The three main plexuses are:

1. The **cervical plexus**, which supplies motor impulses to the muscles of the neck and receives sensory impulses from the neck and the back of the head. The cervical plexus also gives off the phrenic nerve, which activates the diaphragm.
2. The **brachial** (bra'ke-al) **plexus**, which sends numerous branches to the shoulder, the arm, the forearm, the wrist and the hand.
3. The **lumbosacral** (lum-bo-sa'kral) **plexus**, which supplies nerves to the lower extremities. The largest of these branches is the **sciatic** (si-at'ik) nerve, which leaves the dorsal part of the pelvis and extends down the back of the thigh. At its beginning it is nearly an inch thick, but it soon sends branches to the thigh muscles; and near the knee it divides into two subdivisions that supply the leg and the foot.

The Autonomic Nervous System

PARTS OF THE AUTONOMIC NERVOUS SYSTEM

Although the internal organs such as the heart, the lungs, and the stomach contain sensory nerve endings and nerve fibers for conducting sensory messages to the brain and cord, most of these impulses do not reach consciousness. These sensory impulses from the viscera, as is the case of many from the skin and the muscles, are translated into reflex responses without reaching the higher centers of the brain. The sensory neurons from the organs are grouped with those that come from the skin and voluntary muscles. On the other hand, the efferent neurons that supply the glands and the involuntary muscles are arranged very differently from those that supply the voluntary muscles. This variation in the location and arrangement of the **visceral efferent** neurons has led to their classification as part of a separate division called the autonomic nervous system (Fig. 7.12).

The autonomic nervous system has many special parts including ganglia that serve as relay stations. In these ganglia each message is transferred (at a point of contact or **synapse**) from the first neuron to a second one which then carries the impulse to the muscle or gland cell. In the case of voluntary muscle cells each nerve fiber extends all the way from the spinal cord to the muscle with no intervening relay station. The location of parts of the autonomic nervous system is roughly this:

1. The sympathetic pathway begins in the spinal cord with cell bodies of the **thoracolumbar** (tho-rah-ko-lum'bar) area, which is in the region of the lower neck and chest. From this part of the cord nerve fibers extend to the ganglia of one of the sympathetic trunks. These trunks are two cordlike strands that extend up and down on either side of the spinal column from the lower neck to the upper abdominal region. The beadlike enlargements of this trunk are called the **lateral ganglia**. These ganglia contain the cell bodies of the second set of neurons whose fibers then extend to the glands and involuntary muscle tissues.
2. The parasympathetic pathway begins in the cell bodies of the midbrain, medulla and lower (sacral) part of the spinal cord. From these centers the first set of fibers extends to autonomic ganglia that are usually located near or within the walls of the organs. The pathway then continues along a second set of neurons that stimulate the visceral tissues.

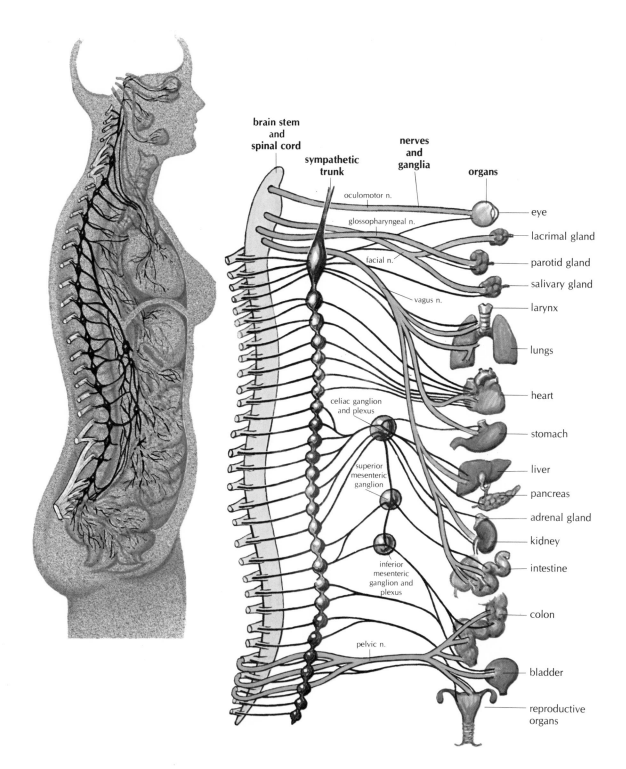

brain stem and spinal cord

sympathetic trunk

nerves and ganglia

organs

oculomotor n.

glossopharyngeal n.

facial n.

vagus n.

celiac ganglion and plexus

superior mesenteric ganglion

inferior mesenteric ganglion and plexus

pelvic n.

eye

lacrimal gland

parotid gland

salivary gland

larynx

lungs

heart

stomach

liver

pancreas

adrenal gland

kidney

intestine

colon

bladder

reproductive organs

Figure 7.12. Topography of the autonomic nervous system.

FUNCTIONS OF THE AUTONOMIC
NERVOUS SYSTEM

The autonomic nervous system regulates the action of the glands, the smooth muscles of hollow organs and the heart. These actions are all carried on automatically; and whenever any changes occur which call for a regulatory adjustment, this is done without our being conscious of it. The sympathetic part of the autonomic nervous system tends to act largely as an accelerator, particularly under conditions of stress. If you will think of what happens to a person who is frightened or angry, you can easily remember the effects of impulses from the sympathetic nervous system:

1. Stimulation of the adrenal gland, which produces hormones, including **epinephrine** (ep-e-nef′rin), that prepare the body to meet emergency situations in many ways (see Chapter 15). The sympathetic nerves and hormones from the adrenal reinforce each other.
2. Dilation of the pupil and decrease in focusing ability (for near objects).
3. Increase in the rate and forcefulness of heart contractions.

4. Increase in blood pressure due partly to the more effective heartbeat and partly to constriction of small arteries in the skin and the internal organs.
5. Dilation of the bronchial tubes in order to allow for more oxygen to enter.
6. Inhibition of peristalsis and of secretory activity so that digestion is slowed.

If you have tried to eat while you were angry, you may have noted that the saliva was thicker and so small in amount that the food was swallowed with difficulty. Then when the food does reach the stomach, it seems to remain there longer than usual.

Once the crisis has passed, the parasympathetic part of the autonomic nervous system normally acts as a balance for the sympathetic system. The parasympathetic system brings about constriction of the pupil, slowing of the heart rate and constriction of the bronchial tubes; and, most important to remember, this system stimulates peristalsis and increases the quantity and fluidity of secretions. The saliva, for example, flows more easily and profusely.

Summary

1. **Nervous system as a whole.**
   A. Function: coordinating system of the body.
   B. Divisions.
      (1) Central nervous system: brain, spinal cord.
      (2) Peripheral nervous system: cranial and spinal nerves.
      (3) Autonomic nervous system (functional classification of a certain group of peripheral nerves).
2. **Nerves:** bundles of nerve fibers, carrying impulses. Nerve tissue is made of nerve cells (neurons) whose components are: cell body, nerve fibers (axons, dendrites). Afferent nerves to central nervous system; efferent nerves from central nervous system.
3. **Brain.**
   A. Main parts: cerebrum (2 hemispheres), midbrain, pons, medulla, cerebellum.
   B. Fissures (sulci): longitudinal fissure, central fissure, lateral fissure.
   C. Ventricles filled with cerebrospinal fluid.
   D. Cerebral cortex: highest functions of brain performed here.
      (1) Frontal lobe (motor cortex).
      (2) Parietal lobe (sensory area).
      (3) Temporal lobe (auditory center).
      (4) Occipital lobe (visual area).
   E. Speech centers: auditory, visual, motor, written.
   F. Other parts of cerebral hemispheres: internal capsule, basal ganglia, olfactory area.
   G. Interbrain (thalamus and hypothalamus) and midbrain.
   H. Cerebellum: aids in muscle coordination, balance, muscle tone.
   I. Pons: links parts of brain; some reflex action.
   J. Medulla: contains respiratory, cardiac, vasomotor centers.
   K. Ventricles of the brain.
   L. Brain waves.

4. **Spinal cord.**
   A. Structure: H-shaped gray matter, surrounded by white matter; all inside spinal canal.
   B. Function: reflexes, conducts sensory impulses to brain, conducts motor impulses from brain to organs. Reflex pathway: nerve cells are sensory (outside cord in ganglia), central, motor.

5. **Coverings of brain and spinal cord.**
   A. Meninges: dura mater, arachnoid, pia mater.

6. **Cerebrospinal fluid.**
   A. Cushions shocks.
   B. Constant formation, flow and absorption.

7. **Cranial nerves.**
   A. General functions: special sense impulses, general sense impulses, voluntary muscle control, involuntary control.
   B. Names: olfactory, optic, oculomotor, trochlear, trigeminal, abducens, facial, acoustic, glossopharyngeal, vagus, accessory, hypoglossal.

8. **Spinal nerves.**
   A. Attached to spinal cord by dorsal and ventral roots. Dorsal roots conduct sensory impulses, ventral roots supply muscles and glands.
   B. Branches (plexuses): cervical, brachial, lumbosacral.

9. **Autonomic nervous system.**
   A. Regulates action of glands, smooth muscles, heart.
   B. Divisions.
      (1) Sympathetic nervous system: origin in thoracolumbar area. Accelerates some body processes.
      (2) Parasympathetic nervous system: some originate in cranial, others in spinal (sacral) region. Balances action of sympathetic system.

Questions and Problems

1. What is the main function of the nervous system?
2. Name the 3 divisions of the nervous system.
3. Describe a neuron and name its parts.
4. Name and locate the main parts of the brain and briefly describe the main functions of each.
5. Name 4 divisions of the cerebral cortex and state what each does.
6. Name and describe the 4 speech centers.
7. Describe the thalamus; where is it located? What are its functions?
8. What activities does the hypothalamus regulate?
9. Locate and describe the spinal cord. Name 3 of its functions.
10. Describe a typical reflex action.
11. Name the covering of the brain and the spinal cord and its divisions. What is an infection of this covering called?
12. What is the purpose of the cerebrospinal fluid? Describe its flow.
13. Name 4 general functions of the cranial nerves.
14. Name and describe the functions of the 12 cranial nerves.
15. Locate the spinal nerves and name 3 main branches of each of them.
16. Describe the function of the autonomic nervous system.
17. Name the 2 parts of the autonomic nervous system, and show how they work during and following a moment of extreme fear.

The Sensory System

Senses and Sensory Mechanisms

The word "sense" might be defined as "the interpretation, by the specialized areas of the cerebral cortex, of an impulse arising from the receptors which are designed to report changes taking place either within the body or outside of it." These sensory receptors consist of the endings of the dendrites of afferent neurons. Some receptors are designed to respond only to special stimuli (sound waves, light rays) while others respond to such general sensations as pain or pressure.

The senses have been said to number five. Actually there are more than that. A partial list includes the following:

1. Visual sense from the eye.
2. Hearing sense from the ear.
3. Taste sense from the tongue receptors.
4. Smell sense from the upper nasal cavities.
5. Pressure, heat, cold, pain and touch from the skin.
6. Position and balance sense from the muscles, the joints and the semicircular canals in the ear.
7. Hunger and thirst senses from various internal parts of the body.

The Eye

PROTECTION OF THE EYEBALL AND ITS PARTS

In the embryo the eye develops as an outpocketing of the brain. As in the case of the brain, the eye is a delicate organ. Therefore, nature has carefully protected the eye by means of the following:

1. The skull bones that form the eye **orbit** (cavity) serve to protect more than half of the eyeball at the back (dorsally).
2. The lids and the eyelashes aid in protecting the eye at the front part (anteriorly).
3. The tears wash away small foreign objects that may enter the lid area.
4. A sac lined with an epithelial membrane separates the front of the eye from the eyeball proper and aids in the destruction of some of the pathogenic bacteria that may enter from the outside.

COATS OF THE EYEBALL

The eyeball has three separate coats or tunics. The outermost layer is called the **sclera** (skle'rah) and is made of firm, tough, connective tissue. It is commonly referred to as the white of the eye. The second tunic of the eyeball is known as the **choroid** (ko'roid) **coat**. It is heavily pigmented, preventing light rays from scattering and reflecting (bouncing) off the inner surface of the eye. The choroid coat might be compared to the dull black lining of a camera. It is made of a delicate network of connective tissue containing much dark brown pigment and interlaced with many blood vessels. The innermost coat, called the **retina** (ret'i-nah), includes some ten different layers of nerve cells, including the end organs commonly called the **rods** and **cones**. These are the receptors for the sense of vision. The rods are sensitive to white and black. The cones are sensitive to color. As far as is known, there are three types of cones, each of which is sensitive to one of the primary colors (red, yellow and blue). Persons who completely lack cones are totally color blind; those who lack one type of cone are partially color blind. Color blindness is an inherited condition and occurs almost exclusively in males (Fig. 8.1).

PATHWAY OF LIGHT RAYS

Light rays pass through a series of transparent, colorless eye parts. On the way they undergo a process of bending known as **refraction**. This refracting of the light rays makes it possible for light from a very large area to be focused upon a very small surface, the retina, where the receptors are located. The following are, in order from outside in, the transparent refracting parts, or **media**, of the eye:

1. The **cornea** (kor'ne-ah) is a forward continuation of the outer coat, but it is transparent and colorless, whereas the sclera is opaque and white.
2. The **aqueous humor**, a watery fluid which fills much of the eyeball in front of the lens, helps to maintain the slight forward curve in the cornea.
3. The **crystalline lens** is a circular structure made of a jellylike material.
4. The **vitreous body** fills the entire space behind the lens and is necessary to keep the eyeball in its spherical shape.

The cornea is referred to frequently as the "window" of the eye. It bulges forward slightly and is the most important refracting structure. Injuries caused by foreign objects or by infection may result in scar formation in the cornea and a resulting area of opacity through which light rays cannot pass. If such an injury involves the central area in front of the pupil (the hole in the center of the colored part of the eye), blindness may be the result. The cornea may be transplanted; eye banks store corneas obtained from donors immediately after death, or in some cases, before death.

The next light-bending medium is the aqueous humor, followed by the crystalline lens. The lens has two bulging surfaces, so it may be best described as biconvex. During youth the lens is elastic and therefore is an important part of the system of accommodation to near vision. In the process of accommodation the lens becomes thicker and thus bends the light rays a greater amount, as is required for near objects. With aging the lens loses its elasticity, and therefore its ability to adjust by thickening, resulting in what is known as the **old eye**, or presbyopia (pres-be-o'pe-ah).

The last of these transparent refracting parts of the eye is the vitreous body. As in the case of the aqueous humor it is important in maintaining the ball-like shape of the eyeball as well as aiding in refraction. The vitreous body is not replaceable; an injury that causes a loss of an appreciable amount of the jellylike vitreous material will cause collapse of the eyeball. This will require the removal of the eyeball, an operation called **enucleation** (e-nu-kle-a'shun).

MUSCLES OF THE EYE

Certain muscles are inside the eyeball itself, and therefore are described as **intrinsic** (in-trin'sik), while others are attached to bones of the eye orbit as well as to the sclera and are called **extrinsic** (eks-trin'sik) muscles.

The intrinsic muscles are found in two circular structures:

1. The **iris**, the colored or pigmented part of the eye, which has a central opening called the pupil. The size of the pupil is governed by the action of two sets of muscles, one of which is arranged

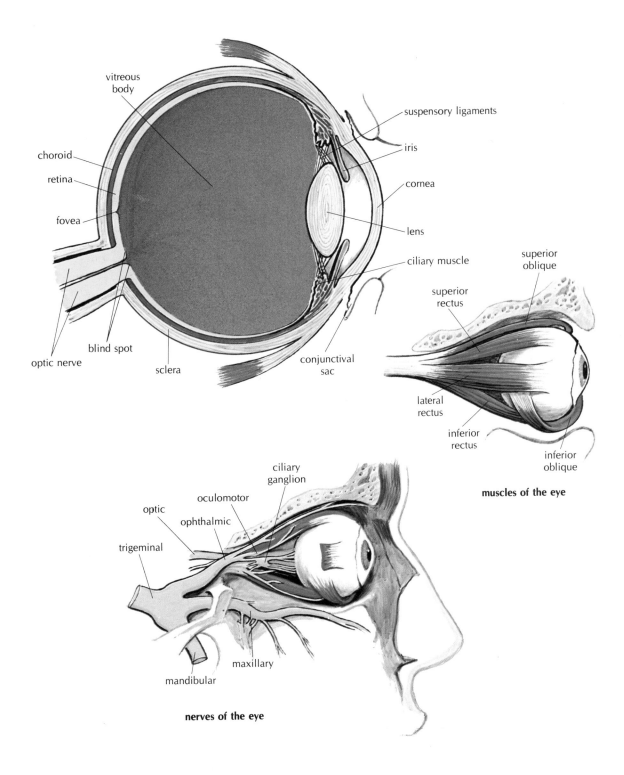

Figure 8.1. The eye.

in a circular fashion, while the other extends in a radial manner resembling the spokes of a wheel.

2. The **ciliary body**, which is shaped somewhat like a flattened ring with a hole that is the size of the outer edge of the iris. This muscle alters the shape of the lens.

The purpose of the iris is to regulate the amount of light entering the eye. If a strong light is flashed in the eye, the circular muscle fibers of the iris, which form a sphincter, contract and thus reduce the size of the pupil. On the other hand, if the light is very dim, the radial involuntary iris muscles, which are attached at the outer edge, contract; the opening is pulled outward and thus enlarged. This pupillary enlargement is known as **dilation** (di-la'shun).

The pupil changes size, too, according to whether one is looking at a near object or a distant one. Viewing a near object causes the pupil to become smaller; a far view will cause it to enlarge.

The muscle of the ciliary body is similar in direction and method of action to the radial muscle of the iris. When the ciliary muscle contracts, it removes the tension on the suspensory ligament of the lens. The elastic lens then recoils and becomes thicker in much the same way that a rubber band would thicken if a pull on it were released. This action changes the focus of the lens, and thus adjusts the eye for either long views or close-ups.

The six extrinsic muscles connected with each eye are ribbonlike and extend forward from the apex of the orbit behind the eyeball. One end of each muscle is attached to a bone of the skull, while the other end is attached to the white (sclera) of the eye. These muscles pull on the eyeball in a coordinated fashion that causes the two eyes to move together in order to center on one visual field. There is another muscle located within the orbit which is attached to the upper eyelid. When this muscle contracts, it keeps the eye open (Fig. 8.1).

NERVE SUPPLY TO THE EYE

The two sensory nerves of the eye are:

1. The **optic nerve**, which carries visual impulses received by the rods and cones in the retina to the brain. This,

it will be recalled, is the second cranial nerve.

2. The **ophthalmic** (of-thal'mik) **nerve** which carries impulses of pain, touch and temperature from the eye and surrounding parts. It is a branch of the fifth (trigeminal) cranial nerve.

The optic nerve is connected with the eyeball a little toward the medial or nasal side of the eye at the back. At this region there are no rods and cones; and so this part, which is a circular white area, is called the **blind spot**, known also as the **optic disk**. There is a tiny depressed area in the retina called the **fovea centralis** (fo've-ah sen-tra'lis), which is the clearest point of vision.

There are three nerves that carry motor fibers to the muscles of the eyeball. The largest is the **oculomotor** nerve, which supplies motor fibers, voluntary and involuntary, to all the muscles but two. The other two nerves, the **trochlear** and the **abducens**, supply one voluntary muscle each.

THE LACRIMAL APPARATUS AND THE CONJUNCTIVAL SAC

The **lacrimal** gland produces tears and is located above the eye toward one side; that is, it is superior and lateral to the eyeball. Tiny tubes carry the tears to the front surface of the eyeball, where they serve to constantly wash the sac that separates the front part of the eyeball from the larger back portion. This sac is lined with a membrane called the **conjunctiva** (kon-junk-ti'vah). The conjunctiva lines the eyelids and is reflected onto the front of the eyeball. It is kept moist by the tears flowing across the front of the eye. Tears then are carried into tiny openings near the nasal corner of the eye. From these openings tears are carried by tubes that drain into the nose via the nasolacrimal duct. A slight excess of tears, or **lacrimation** (lak-re-ma'shun), causes nose blowing; and a greater overproduction of tears results in an overflow onto the face (Fig. 8.2).

The Ear

The ear is a combination sensory organ, related to both hearing and equilibrium (Fig. 8.3). It may be divided into three main sections namely:

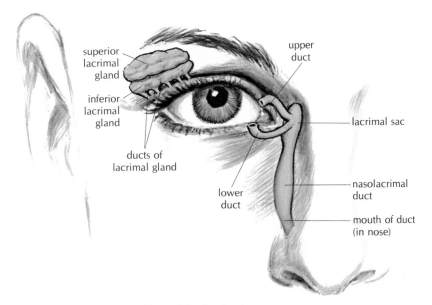

Figure 8.2. Lacrimal apparatus.

1. The **external ear**, which includes the outer projection and a canal.
2. The **middle ear**, which is an air space containing three small bones.
3. The **internal ear**, which is the most important part, since it contains the sensory end organs or receptors for hearing and equilibrium.

THE EXTERNAL EAR

The projecting part of the ear is known as the **pinna** (pin'nah), or the **auricle** (aw're-kl). From a functional point of view it is probably of little importance in the human. Then follows the opening itself, the **external auditory canal**, which extends medially for about 1 inch or more, depending upon which wall of the canal is measured. The skin lining this tube is very thin, and in the first part of the canal contains many wax, or **ceruminous** (se-roo'me-nus), glands. The **cerumen** (se-roo'men), or wax, may become dried and impacted in the canal so that removal by a physician is required. The same kinds of disorders that involve the skin elsewhere also may affect the skin of the external auditory canal: eczema, boils and other infections.

At the end of the auditory canal is the **tympanic** (tim-pan'ik) **membrane**, or eardrum. It serves as a boundary between the external auditory canal, or **meatus** (me-a'tus), and the middle ear cavity. It may be injured by inserted objects such as bobby pins or toothpicks. Normally the air pressure on the two sides of the drum is equalized by means of the **eustachian** (u-sta'ke-an) **tube** connecting the middle ear cavity and the throat (pharynx), allowing the eardrum to vibrate freely with the incoming sound waves. Sudden great changes in the pressure on either side of the eardrum may cause excessive stretching and inflammation of the membrane. There may even be perforation of the drum. In some cases pressure from pus or exudate in the middle ear cavity can be relieved only by cutting the eardrum, a procedure called a **myringotomy** (mir-in-got'o-me).

THE MIDDLE EAR

The middle ear cavity is a small flattened space that contains air and three small bones, or **ossicles** (os'e-kls). Air is brought into the cavity through the eustachian tube (also called the **auditory tube**; it should not be confused with the external auditory canal). The eustachian tube connects the lower part of the middle ear cavity with the pharynx. The mucous membrane of the pharynx is continuous through the eustachian tube into the middle ear cavity, and infection travels along the membrane, causing middle ear disease. This happens more often in children, partly because the tube is more horizontal in the child, while in the adult the tube tends to

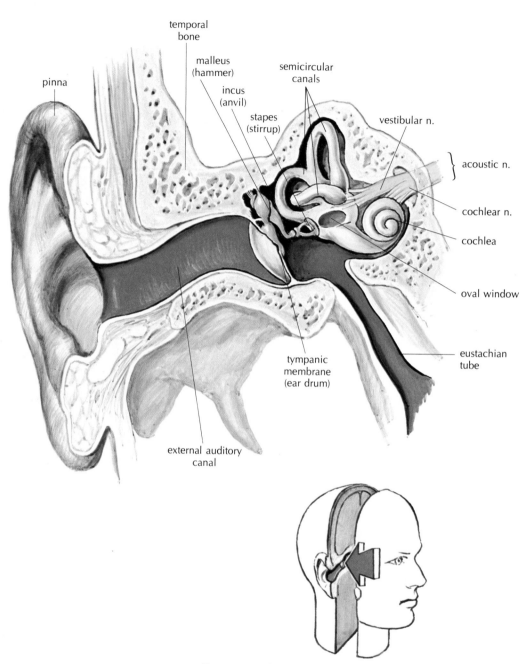

Figure 8.3. The ear.

slant toward the pharynx. At the back of the middle ear cavity is an opening into the **mastoid air cells**, which are spaces inside a part of the temporal bone, one of the major bones of the skull.

The three ossicles are joined in such a way that they amplify the sound waves received by the tympanic membrane and then transmit the sounds to the fluid in the internal ear. The handlelike part of the first bone, or **malleus** (mal'e-us), is attached to the tympanic membrane, while the headlike portion connects with the second bone, which is called the **incus** (ing'kus). The innermost of the ossicles is shaped somewhat like a stirrup and is called the **stapes** (sta'pez). It is connected with the membrane of the **oval window** which in turn vibrates and conducts these waves to the fluid of the internal ear.

THE INTERNAL EAR

The most complicated and important part of the ear is the internal portion. It includes three separate spaces hollowed out inside the temporal bone. Because they are rather complex, they constitute what has been called the **bony labyrinth** (lab'i-rinth). Next to the oval window is the **vestibule**. This entrance area then communicates with the bony tube shaped like a snail shell, called the **cochlea** (kok'le-ah), toward the front, and with the **semicircular canals** toward the back. These spaces all contain a fluid called **perilymph**. In the fluid of the bony semicircular canals are the **membranous** (mem'brah-nus) **canals**, which contain another fluid called **endolymph**. In a similar fashion a **membranous cochlea** is situated in the perilymph of the bony cochlea, and it also is filled with endolymph. The organ of hearing, made of receptors connected with nerve fibers in the **cochlear nerve** (a part of the acoustic nerve), is located inside the membranous cochlea, or **cochlear duct**. The sound waves enter the external auditory canal and cause the eardrum to vibrate. These vibrations are amplified by the ossicles and transmitted by them to the perilymph. They then are conducted by the perilymph through the membrane to the endolymph. The waves of the endolymph are transmitted to the tiny hairlike receptors, which are stimulated and conduct nerve impulses through the nerve fibers to the brain for interpretation (Fig. 8.4).

The semicircular canals contain the sensory organs related to equilibrium. The membranous canals are connected with two small sacs in the vestibule, and one of these sacs contains sensory end organs for obtaining information with relation to the position of the head. Nerve fibers from these sacs and from the canals form the **vestibular** (ves-tib'u-lar) **nerve** which joins the cochlear nerve to form the acoustic nerve, which latter, as we learned, is one of the 12 cranial nerves.

Other Organs of Special Sense

TASTE SENSE

The sense of taste involves receptors in the tongue and two different nerves that carry taste impulses to the brain. The taste receptors are known as **taste buds** and are located along the edges of small depressed areas called **fissures**. These taste buds are stimulated only if the substance to be tasted is in solution. Tastes have been described as essentially of four kinds, namely:

1. Sweet tastes, which are most acutely experienced at the tip of the tongue.
2. Sour tastes, most effectively detected by the taste buds located at the sides of the tongue.
3. Salty tastes, which, as in the case of sweet tastes, are most acute at the tip of the tongue.
4. Bitter tastes, which are detected at the back (dorsal) part of the tongue.

The nerves of taste include the facial and the glossopharyngeal nerves. The interpretation of taste impulses probably is accomplished by the lower front portion of the brain, although there may not be a sharply separate taste or **gustatory** (gus'tah-to-re) center.

SENSE OF SMELL

The sensory end organs, or receptors, for smell are located in the olfactory **epithelium** of the upper part of the nasal cavity. Because they are high in the nasal cavity, an animal or a person "sniffs" in order to bring the gases responsible for an odor upward in the nose. The pathway of the impulses from the receptors for smell is the olfactory nerve. This leads to the olfactory center in the brain. The interpretation of smell is closely related to the sense of taste. The smell of foods is just as important in stimulating appetite and the flow of digestive juices as is the sense of taste (Fig. 8.5).

HUNGER AND APPETITE

Hunger includes intermittent sensation coming from the region of the stomach. It is due, in part, to contractions of the stomach muscle, and is not continuous; that is, if a person is starving, the hunger pangs diminish instead of becoming more acute. Appe-

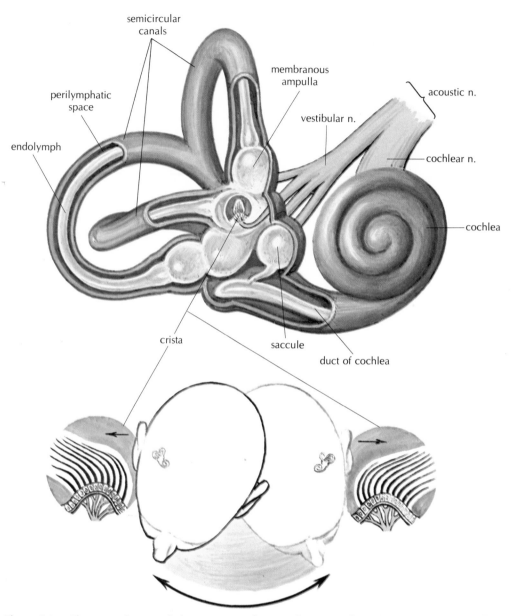

Figure 8.4. The internal ear, including a section showing the crista where the sensory receptors for balance are located.

tite differs from hunger in that although it is basically a desire for food, it often has no relationship to the need for food. Hunger may have been relieved by an adequate meal, but the person may still have an appetite for additional food. A loss of appetite is known as **anorexia** (an-o-rek'se-ah), and may be due to a great variety of physical and mental disorders. The location of the nerve receptors which transmit hunger impulses is still un-

certain. They are probably in the stomach muscles.

SENSE OF THIRST

Although thirst may be due to a generalized lack of water in the tissues, the sense of thirst seems to be largely localized in the mouth, the tongue and the pharynx. It is a very unpleasant sensation and is continuous up to relief or death. If there is an excessive excretion of water, as in diabetes, there may be excessive thirst, which is called **polydipsia** (pol-e-dip′se-ah).

General Senses

As opposed to the **special** senses, in which the receptors are limited to a relatively small area in the body, the **general** senses are scattered throughout the body. These may be said to include pressure, heat, cold, pain, touch, position and balance senses, all of which are rather widely distributed.

PRESSURE SENSE

It has been found that even though the skin is anesthetized, there still is consciousness of pressure. These end organs for deep sensibility are located in the subcutaneous and deeper tissues. They are sometimes referred to as receptors for deep touch.

TEMPERATURE SENSE

Heat and cold receptors have separate nerve fiber connections. Each has its type of end organ structure peculiar to it, and the distribution of each varies considerably. A warm object will stimulate only the heat receptors, while a cool object affects only the cold terminals. More heat receptors are found in the lips than in the hands, so that they are more sensitive to heat than are the hands. As in the case of other sensory receptors, continued stimulation results in **adaptation**; that is, the receptors adjust themselves in such a way that one does not feel a sensation so acutely if the original stimulus is continued. For example, the initial immersion of a hand in hot water may give rise to an uncomfortable sensation; however, if the immersion is prolonged, the water very soon will not feel as hot as it did at first (even if it has not cooled appreciably).

SENSE OF TOUCH

The touch receptors are small rounded bodies called **tactile** (tak′til) **corpuscles.** They are found mostly in the dermis and are especially close together in the tips of the fingers and the toes. The tip of the tongue also contains many of these receptors and so is very sensitive to touch, whereas the back of the neck is relatively insensitive.

PAIN SENSE

Pain is the most important protective sense. The receptors for pain are the most widely distributed sensory end organs. They are found in the skin, the muscles and the joints, and to a lesser extent in most internal organs (including the blood vessels and viscera). Pain receptors are not oval bodies as are many of the other sensory end organs, but apparently are merely branchings of the nerve fiber, called **free nerve endings.** **Referred pain** is a term used in cases in which pain that seems to be in an outer part of the body, particularly the skin, actually originates in an internal organ located near that particular area of skin. These areas of referred pain have been mapped out on the basis of much experience and many experiments. It has been found, for example, that liver and gallbladder disease often cause referred pain in the skin over the right shoulder. Spasm of the coronary arteries that supply the heart may cause pain in the left shoulder and the left arm. One reason for this is that some neurons have the twofold duty of conducting impulses both from visceral pain receptors and from pain receptors in neighboring areas of the skin. The brain cannot differentiate between these two possible sources; but since most pain sensations originate in the skin, the brain automatically assigns the pain to this more likely place of origin.

Pain sense differs from other senses in that continued stimulation does not result in adaptation. This is nature's way of being certain that the warnings of the pain sense are heeded. Sometimes the cause cannot be remedied quickly, and occasionally not at all. Then it is necessary to relieve pain.

SENSE OF POSITION

Receptors located in muscles, tendons and joints relay impulses that aid in judging the position and changes in the locations of parts with respect to each other. They also inform the brain of the amount of muscle contraction and tendon tension. These rather widely spread end organs, which are known as **proprioceptors** (pro-pre-o-sep′tors), are aided in this function by the semicircular

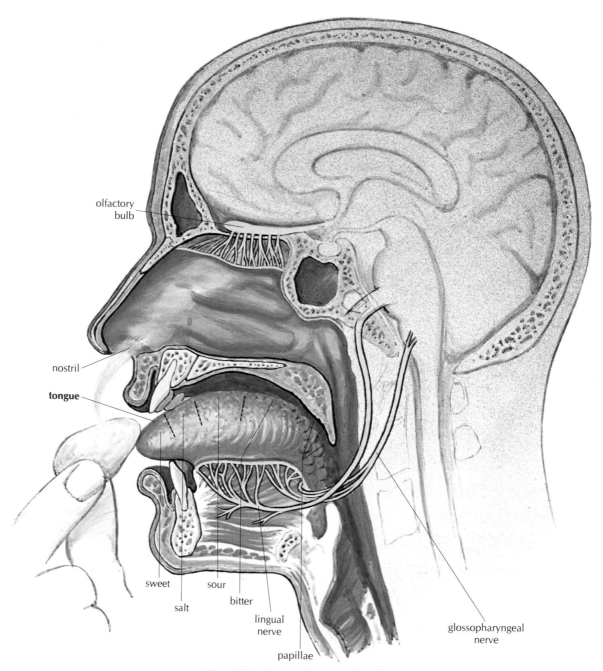

Figure 8.5. Organs of taste and smell.

canals and related internal ear structures. Information received by these receptors is needed for coordination of muscles and is important in such activities as walking, running and many more complicated skills such as playing a musical instrument. These muscle sense end organs also play an important part in maintaining muscle tone and good posture, as well as allowing for the adjustment of the muscles for the particular kind of work to be done. The nerve fibers that carry impulses from these receptors enter the spinal cord and ascend to the brain in the back (posterior) part of the cord.

Summary

1. **Senses:** sight, hearing, taste, smell, pressure, heat, cold, pain, touch, position, balance, hunger, thirst.
2. **Eye.**
   A. Parts and purposes.
      (1) Protection: orbits, lids, eyelashes, tears, epithelial sac.
      (2) Coats: sclera, choroid, retina.
      (3) Light path: cornea, aqueous humor, lens, vitreous body.
      (4) Muscles: intrinsic (iris, ciliary body); 6 extrinsic.
      (5) Nerves: optic (visual impulses from rods and cones of retina); ophthalmic (pain, touch, temperature impulses from eye and surrounding parts); 3 motor nerves.
      (6) Lacrimal apparatus: lacrimal gland produces tears which moisten conjunctiva.
3. **Ear.**
   A. Parts and purposes.
      (1) Divisions: external, middle, internal.
      (2) External: pinna, auditory canal, tympanic membrane.
      (3) Middle: ossicles (malleus, incus, stapes) amplify sounds from tympanic membrane, transmit them to oval window. Eustachian tube connects to pharynx, equalizes pressure, pathway for infection.
      (4) Internal: bony labyrinth. Oval window, vestibule, cochlea, semicircular canals, all contain perilymph. Membranous canals (in semicircular canals), membranous cochlea (in cochlea) both filled with endolymph. Receptors in cochlear duct make up the organ of hearing.
      (5) Path for sound: eardrum vibrates, vibrations amplified by ossicles, transmitted to perilymph, to endolymph, to nerve receptors, to nerves, to brain.
      (6) Equilibrium: membranous canals connected with 2 sacs, 1 sac containing sensory nerves indicating position of head.
4. **Other special sense organs.**
   A. Taste: receptors (taste buds on tongue). Four tastes (sweet, sour, salty, bitter).
   B. Smell: receptors (olfactory epithelium of nasal cavity).
   C. Hunger and appetite: hunger due to stomach muscle contractions; not continuous; appetite (desire for food). Hunger receptors probably in stomach muscles.
   D. Thirst: receptors in mouth, tongue, throat; continuous.
5. **General senses.**
   A. Pressure: end organs in deep tissues.
   B. Temperature: heat and cold receptors separate. Adaptation (common to most other senses also).
   C. Touch: receptors (tactile corpuscles). Close together in fingers, toes, tongue.
   D. Pain: protective, no adaptation. Referred pain (from deeper organs but seemingly originating in nearby skin area. Areas mapped out for diagnostic purposes).
   E. Position: receptors (proprioceptors in muscles, tendons, joints aided by semicircular canals). Nerve fibers enter spinal cord; ascend to the brain.

Questions and Problems

1. Give a general definition of a sense and name 7 of the senses.
2. Name the main parts of the eye and trace the path of a light ray from the outside of the eye to the brain. Show the action of muscles.
3. Outline the main parts of the ear and describe the process that ensues from the time that a sound wave activates the eardrum to the registration of the sound in the brain.
4. Name the 4 kinds of taste. Where are the taste receptors?
5. Describe the olfactory apparatus.
6. What is the difference between hunger and appetite?
7. What is the difference between a general and a special sense?
8. What does "adaptation" mean, with respect to the senses? Does this occur in the case of every sense?
9. Explain referred pain and give an example of its occurrence.
10. Where are the receptors for the senses of position and balance (equilibrium) located?

The Heart

Circulation and the Heart

In the next two chapters we shall investigate the manner in which the blood acquires its food and oxygen to be delivered to the cells, and disposes of the waste products of cell metabolism. This continuous one-way movement of the blood is known as its **circulation**. The fact that blood circulates throughout the body implies that there must be some sort of propelling mechanism. The prime mover in this case is the **heart**; and we shall have a look at the heart before going into the circulatory vessels in any detail.

The heart is a muscular pump which drives the blood through the blood vessels. This organ is slightly bigger than a fist, and is located between the lungs in the center and a bit to the left of the midline of the body. The strokes (contractions) of this pump average about 72 per minute and are carried on unceasingly for the whole of a lifetime (Fig. 9.1).

The importance of the heart has been recognized for centuries. The fact that its rate of beating is affected by the emotions may be responsible for the very frequent references to the heart in song and poetry. However, the vital functions of the heart and the tragic increase in heart disease are of more practical importance to us at this time.

Structure of the Heart

LAYERS OF THE HEART WALL

The heart is a hollow organ the walls of which are formed of three different layers. Just as a warm coat might have a smooth lining, a thick and bulky interlining and an outer layer of a third fabric, so the heart wall has three tissue layers, as follows:

1. **Endocardium** (en-do-kar'de-um), a very smooth layer of cells that resembles squamous epithelium. This membrane lines the interior of the heart, and is also the material of which the valves of the heart are formed.
2. **Myocardium** (mi-o-kar'de-um), which is the muscle of the heart and is much the thickest layer.
3. **Pericardium** (per-i-kar'de-um), which forms the outermost layer of the heart wall as well as serving as the lining of the pericardial sac (see Chapter 2).

TWO HEARTS AND A PARTITION

Physicians often refer to the right heart and the left heart. This is because the human heart is really a double pump. The two sides are completely separated from each other by a partition called the **septum**. The upper part of this partition is called the **interatrial** (in-ter-a'tre-al) **septum**, while the

109

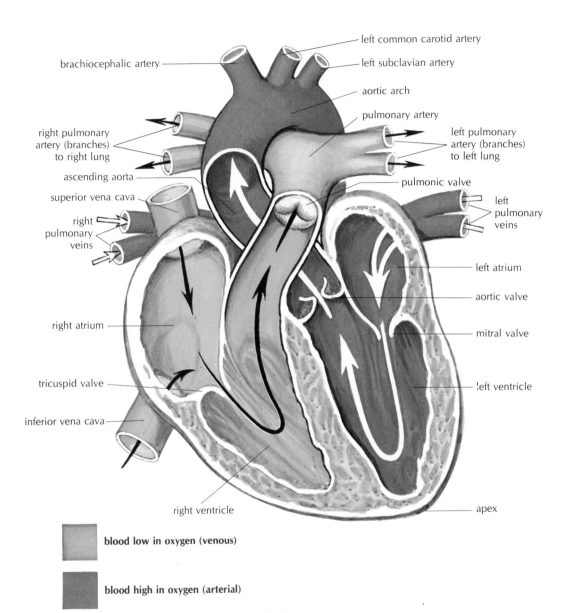

left common carotid artery

left subclavian artery

brachiocephalic artery

aortic arch

pulmonary artery

right pulmonary artery (branches) to right lung

left pulmonary artery (branches) to left lung

ascending aorta

pulmonic valve

superior vena cava

left pulmonary veins

right pulmonary veins

left atrium

right atrium

aortic valve

mitral valve

tricuspid valve

left ventricle

inferior vena cava

right ventricle

apex

■ blood low in oxygen (venous)

■ blood high in oxygen (arterial)

Figure 9.1. The heart and the great vessels.

larger lower portion is called the **interventricular** (in-ter-ven-trik'u-lar) **septum**. This septum, as in the case of the heart wall, is largely myocardium.

FOUR CHAMBERS

On either side of the heart there are two chambers, one of which is a receiving space and the other a pumping chamber. These four chambers are called:

1. The **right atrium**, which is a thin-walled space that receives the venous blood returning from the body tissues. This blood is carried in the **veins**, which are the blood vessels leading *to* the heart from the body tissues.

2. The **right ventricle**, which pumps the venous blood dropped into it from the right atrium, and sends it to the lungs.

3. The **left atrium**, which receives blood high in oxygen content as it returns from the lungs.

4. The **left ventricle**, which has the thickest walls of all in order to pump oxygenated blood to all parts of the body. This blood goes through the **arteries**, which is the name for the vessels that take blood *from* the heart to the tissues.

FOUR VALVES

Since the ventricles are the pumping chambers, the valves, which are all one-way, are located at the entrance and the exit of each ventricle (Fig. 9.2). The valves at the entrances are the **atrioventricular** (a-tre-o-ven-trik'u-lar) **valves**, while the exit valves are **semilunar** (sem-e-lu'nar) **valves**. "Semilunar" means "resembling a half-moon." Each valve has a specific name, as follows:

1. The **tricuspid** (tri-kus'pid) **valve**. ("Tricuspid" means "three-pointed." Cusps are the flaps of the valves, which open and close). It closes at the time the right ventricle begins pumping in order to prevent any blood from going back into the right atrium. This is the right atrioventricular valve.
2. The **pulmonary** (pul'mo-nar-e) **semilunar valve**, located between the right ventricle and the pulmonary artery, which leads to the lungs. As soon as the right ventricle has finished emptying itself, the valve closes in order to prevent blood on its way to the lungs from returning to the ventricle.
3. The **mitral** (mi'tral) **valve**, or left atrioventricular valve, which is made of two rather heavy flaps or cusps. This valve closes at the time the

powerful left ventricle begins its contraction. It prevents the blood from returning to the left atrium.
4. The **aortic** (a-or'tik) **semilunar valve** located between the left ventricle and the largest artery, the **aorta** (a-or'tah), prevents the return of aortic blood to the left ventricle.

Physiology of the Heart

THE WORK OF THE HEART

Although the right and the left sides of the heart are completely separated from each other, they work together. The blood is squeezed through the chambers by a contraction of heart muscle beginning in the thin-walled upper chambers, the atria, and followed by a contraction of the thick muscle of the lower chambers, the ventricles. This active phase is called **systole** (sis'to-le), and in each case it is followed by a short resting period known as **diastole** (di-as'to-le). The contraction of the walls of the atria is completed at the time the contraction of the ventricles begins. Thus the resting phase (diastole) begins in the atria at the same time as the contraction (systole) begins in the ventricles. As soon as the ventricles have emptied, the atria (which meanwhile have been filling with blood) contract while the ventricles relax and again fill with blood. Then the ventricular systole begins (Fig. 9.3).

CONTROL OF THE HEARTBEAT

If the nerves which supply the voluntary muscles are cut, these muscles cease to function; that is, they are completely para-

coronary arteries

aortic valve   tricuspid valve   mitral valve

Figure 9.2. Valves of the heart.

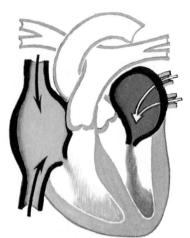
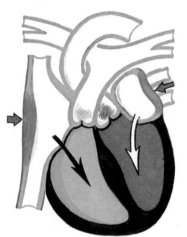

diastole phase
right and left atria fill with
blood; blood flows into
ventricles;

atria contract; blood
is squeezed into ventricles

systole phase
contracting ventricles
squeeze blood into aorta
and pulmonary arteries

Figure 9.3. Pumping cycle of the heart.

lyzed. If the nerves which supply the heart
are severed, however, the heart will continue to beat. The reason for this is that
although the heart is under the control of
the nervous system, heart muscle itself is
capable of contracting rhythmically independently of outside control. Despite this
property of **automaticity** (au-to-mah-tis′e-te),
the impulses from the nervous system are
required to cause a rapid enough beat to
maintain circulation effectively. Without
nerve connection the heart rate might be
less than 40 beats per minute instead of the
usual 70 to 90 per minute.

THE CONDUCTION SYSTEM OF THE HEART

Specialized masses of tissue in the heart
wall form the conduction system of the
heart, regulating the order of events. Two
of these are called **nodes**, while the third is a
branching structure called the **atrioventricular bundle**. The **sinoatrial node** is located in
the upper wall of the right atrium and acts
as a pacemaker. The second node is called
the **atrioventricular node** and is located in
the septum at the junction between the
interatrial portion and the interventricular part (see Fig. 9.4). The atrioventricular
bundle, which is also known as the **bundle
of His**, is located in the interventricular septum with branches extending to all parts of

the ventricle walls. The order in which the
impulses travel is as follows:

1. The beginning of the heartbeat is in
   the sinoatrial node, the pacemaker.
2. The excitation (contraction) wave
   travels throughout the muscle of the
   atria, causing them to contract.
3. The atrioventricular node is stimulated next, and transmits the wave to
   the bundle of His, with a rapid spread
   to all parts of the ventricle walls.
4. The entire ventricular musculature
   contracts practically all at once.

HEART SOUNDS AND MURMURS

The normal heart sounds are usually
described by the two syllables "lubb" and
"dupp." The first is a longer and lower-
pitched sound which occurs during the ventricular systole. It is probably caused by a
combination of sounds made by the muscle
of the ventricles and the closure of the atrioventricular valves. The second, or "dupp,"
sound is shorter and sharper. It occurs during the beginning of ventricular relaxation,
and is due in large part to the sudden
closure of the semilunar valves. Abnormal
sounds are called **murmurs** and are due to
faulty action of the valves. If, for example,
the valves fail to close tightly and blood
leaks back, a murmur is heard. Another condition giving rise to an abnormal sound is

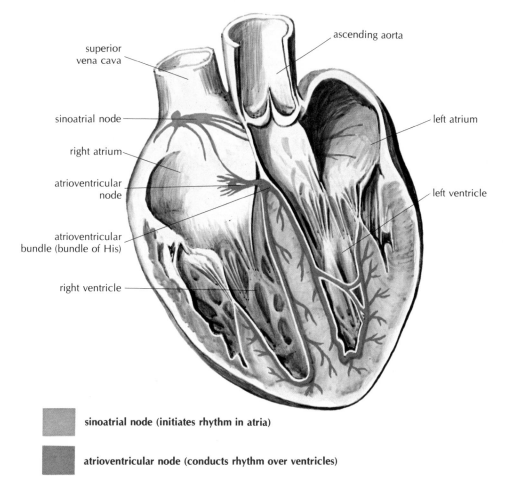

superior vena cava

ascending aorta

sinoatrial node

left atrium

right atrium

atrioventricular node

left ventricle

atrioventricular bundle (bundle of His)

right ventricle

sinoatrial node (initiates rhythm in atria)

atrioventricular node (conducts rhythm over ventricles)

Figure 9.4. The conduction system of the heart.

the narrowing (stenosis) of a valve orifice. The many conditions which can cause an abnormal heart sound may be due to congenital defects, to disease, or to physiologic variations. A murmur may be called a **functional murmur**; that is, not necessarily involving an abnormality. On the other hand, an abnormal sound caused by any structural change in the heart or the vessels connected with the heart is called an **organic murmur.**

Summary

1. **Structure of the heart.**
   A. Three layers: endocardium, myocardium and pericardium.
   B. Two separate pumps, with an intervening septum.
   C. Four chambers: left and right atria and the left and right ventricles.
   D. Four valves: tricuspid, pulmonary, mitral and aortic.
2. **Physiology of the heart.**
   A. Cardiac cycle includes contraction (systole) and resting phase (diastole).
   B. Heart contractions self-sustaining, though nervous control essential for adequate circulation.
   C. Events in heart action regulated by a conduction system of tissue: sinoatrial node, atrioventricular node and bundle of His.
   D. Heart sounds and murmurs.
      (1) First and second sounds: lubb, dupp.
      (2) Murmurs caused by faulty valve action; are functional or organic.

Questions and Problems

1. What are the 3 layers of the heart wall?
2. What are the 2 parts of the partition of the heart called? How do they differ from one another?
3. Name each of the chambers of the heart and tell what each does.
4. Name the valves of the heart. Explain the purpose of each valve.
5. Explain systole and diastole and tell how these phases are related to each other in the 4 chambers of the heart.
6. How does the heart's ability to contract differ from that of other muscles? What is required to maintain an effective rate of heart beat?
7. What are the parts of the heart's conduction system called and where are these structures located? Outline the order in which the excitation waves travel.
8. What 2 syllables are used to indicate normal heart sounds, and at what time in the heart cycle can they be heard?
9. What are the 2 kinds of murmurs and how do they differ from each other? Name 2 conditions which give rise to murmurs.

Blood Vessels and Blood Circulation

Blood vessels, together with the four chambers of the heart, form a closed system for the flow of blood; only if there is an injury to some part of the wall of this system does any blood escape (Fig. 10.1). The circulatory system will be easy to understand now that we know what the blood does and where it is supposed to go. If you keep one eye on the diagrams and the other on the text as the vessels are described, a picture of the system as a whole will gradually emerge.

Kinds of Blood Vessels and Their Functions

On the basis of function, blood vessels may be divided into three groups, as follows:

1. **Arteries**, which carry blood from the pumping chambers of the heart (ventricles) out to the organs and other parts of the body.
2. **Veins**, which drain the tissues and the organs, and return the blood to the heart.

3. **Capillaries**, which allow for exchanges between the blood and the body cells, or between the blood and the air in the lung tissues. The capillaries connect the smaller arteries and veins.

Arteries and veins both may be subdivided into two groups or circuits:

1. **Pulmonary** vessels, which are related to the lungs. They include the pulmonary artery and its branches to the lungs, and the veins that drain the lung capillaries. The pulmonary arteries carry blood low in oxygen from the right ventricle, while the pulmonary veins carry blood high in oxygen from the lungs into the left atrium. This circuit concerns itself with eliminating carbon dioxide from the blood and replenishing its supply of oxygen.
2. **Systemic** (sis-tem′ik) arteries and veins, which are related to the rest of the body. This circuit is concerned with supplying food and oxygen to all the tissues of the body and carrying away waste materials from the tissues for disposal.

115

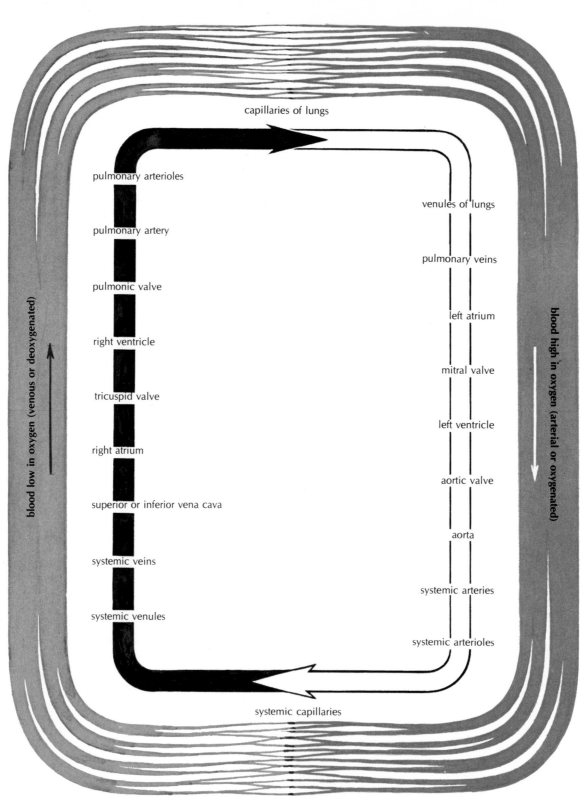

Figure 10.1. Blood vessels constitute a closed system for the flow of blood. Note that changes in oxygen content occur as the blood flows through capillaries.

Structure of Blood Vessels

ARTERY WALLS

The arteries have much the thickest walls because they receive the pumping drive from the ventricles of the heart. There are three coats (tunics) which resemble the three tissue layers of the heart. These are:

1. The innermost membrane of **endothelium**, which forms a smooth surface over which the blood may easily move.
2. The second, more bulky layer, which is made of **involuntary muscle** combined with elastic connective tissue.
3. An outer tunic, which is made of a supporting **connective tissue.**

The largest artery, the **aorta**, is about 1 inch in diameter and has the thickest wall. The smallest subdivisions of arteries, the **arterioles** (ar-te're-oles), have thinner walls in which there is very little connective tissue but relatively more muscle.

CAPILLARY WALLS

The microscopic branches of these tiny connecting vessels have the thinnest walls of any vessels: one cell layer. The capillary walls are transparent and are made of smooth platelike cells that continue from the lining of the arteries. Because of the thinness of these walls, exchanges between the blood and the body cells are possible. The capillary boundaries are the most important center of activity for the entire circulatory system. Their function will be explained later in this chapter.

WALLS OF VEINS

The smallest veins, called **venules** (ven'ules), are formed by the union of capillaries. Their walls are only slightly thicker than those of the capillaries. As the veins become larger, the walls become thicker. However, veins have much thinner walls than those of comparable arteries. Although there are three layers of material in the walls of the larger veins, as in the artery walls, the middle tunic is relatively thin in vein walls. Therefore, veins are easily collapsed, and slight pressure by a tumor or some other mass may interfere with the return blood flow. Most veins are equipped with one-way valves which permit the blood to flow in only one direction. They are most numerous in the veins of the extremities (see Fig. 10.2).

Names of Systemic Arteries

THE AORTA AND ITS PARTS

The aorta is by far the largest artery of the body (Fig. 10.3). It extends upward and to the right from the left ventricle. Then it curves backward and to the left. It continues down behind the heart just in front of the vertebral column, through the diaphragm and into the lower dorsal part of the abdomen. The sections of the aorta are named in much the same manner in which a street is divided into north and south portions. The aorta is one continuous tube divided into the following regions:

1. The **ascending aorta**, which is near the heart and inside the pericardial sac.
2. The **aortic arch**, which curves from the right to the left, and also extends backward.
3. The **thoracic** (tho-ras'ik) **aorta**, which lies just in front of the vertebral column behind the heart and in the space behind the pleura.
4. The **abdominal aorta**, which is the longest section of the aorta, spanning the abdominal cavity.

BRANCHES OF THE ASCENDING AORTA

The first, or ascending, part of the aorta has two branches called the left and right **coronary arteries**, which supply the heart muscle. These form a crown around the base of the heart and give off branches to all parts of the myocardium.

BRANCHES OF THE AORTIC ARCH

The arch of the aorta, located immediately beyond the ascending aorta, sends off three large branches:

1. The **brachiocephalic** (brak-e-o-se-fal'ik) **trunk**, which is a short artery formerly called the innominate. After extending upward for about 2 inches (4 to 5 cm.), it divides into the **right subclavian** (sub-kla've-an) **artery**, which supplies the right upper extremity (arm), and the **right common**

117

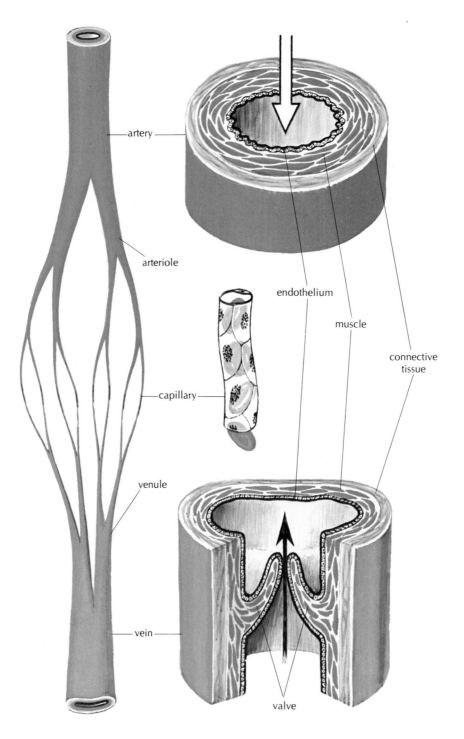

artery

arteriole

endothelium

muscle

connective
tissue

capillary

venule

vein

valve

Figure 10.2. Sections of small blood vessels to show the thicker arterial walls and the thin walls of veins and of capillaries. Venous valves also are shown. The arrows indicate the direction of blood flow.

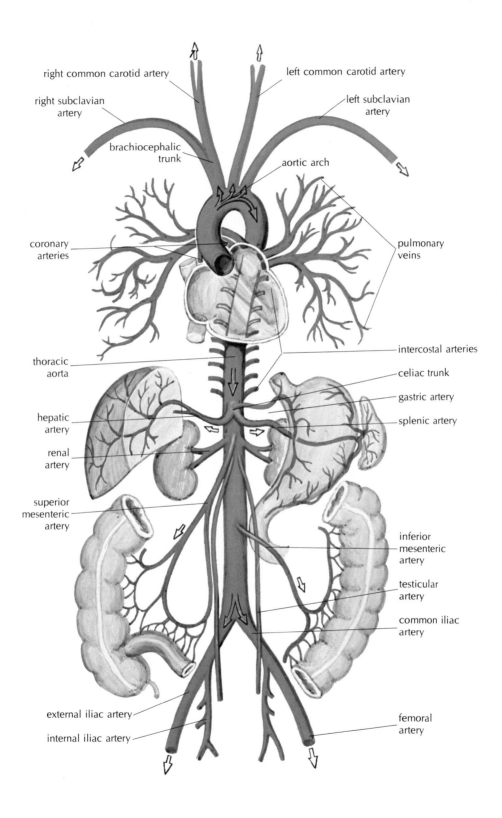

right common carotid artery

left common carotid artery

right subclavian artery

left subclavian artery

brachiocephalic trunk

aortic arch

coronary arteries

pulmonary veins

intercostal arteries

thoracic aorta

celiac trunk

gastric artery

hepatic artery

splenic artery

renal artery

superior mesenteric artery

inferior mesenteric artery

testicular artery

common iliac artery

external iliac artery

femoral artery

internal iliac artery

Figure 10.3. The aorta and its branches. The arrows indicate the flow of blood.

carotid (kah-rot'id) **artery**, which supplies the right side of the head and the neck.

2. The **left common carotid artery**, which extends upward from the highest part of the aortic arch. It supplies the left side of the neck and the head.

3. The **left subclavian artery**, which extends under the left collar bone (clavicle) and supplies the left upper extremity. This is the last branch of the aortic arch.

BRANCHES OF THE THORACIC AORTA

The third part of the aorta supplies branches to the chest wall, to the esophagus (swallowing tube), and to the bronchi (the treelike subdivisions of the windpipe) and their subdivisions in the lungs. There are usually nine to ten pairs of **intercostal** (in-ter-kos'tal) **arteries** that extend between the ribs, sending branches to the muscles and other structures of the chest wall.

BRANCHES OF THE ABDOMINAL AORTA

As in the case of the thoracic aorta, there are unpaired branches extending forward and paired arteries extending toward the side. The unpaired vessels are large arteries that supply the abdominal viscera. The most important of these visceral branches are:

1. The **celiac** (se'le-ak) **trunk**, which is a short artery ½ inch (12 mm.) long that subdivides into three branches, namely, the **left gastric** to the stomach, the **splenic** (splen'ik) to the spleen and the very important **hepatic** (he-pat'ik) **artery** which carries oxygenated blood to the liver.

2. The **superior mesenteric** (mes-en-ter'ik) **artery**, which is the largest of these branches, and which carries blood to most of the small intestine as well as to the first half of the large intestine.

3. The much smaller **inferior mesenteric artery**, which is located lower, near the end of the abdominal aorta, and supplies the last half of the large intestine.

The lateral (paired) branches of the abdominal aorta include the following right and left divisions:

1. The **phrenic** (fren'ik) **arteries**, which supply the diaphragm. The diaphragm is the muscular partition between the abdominal and the thoracic cavities.

2. The **suprarenal** (su-prah-re'nal) **arteries**, which provide blood for the vascular adrenal (suprarenal) glands.

3. The **renal** (re'nal) **arteries**, largest in this group, which carry blood to the kidneys.

4. Arteries that supply the sex glands, called **ovarian arteries** in the female and **testicular** (tes-tik'u-lar) **arteries** in the male (formerly called the spermatic arteries).

5. Four pairs of **lumbar** (lum'bar) **arteries**, which extend into the heavy musculature of the abdominal wall.

ILIAC ARTERIES AND THEIR SUBDIVISIONS

The abdominal aorta finally divides into two common **iliac arteries**. Each of these vessels, about 2 inches (5 cm.) long, extends into the pelvis, where each one subdivides into **internal** and **external** iliac arteries. The internal iliac vessels then send branches to the pelvic organs, including the urinary bladder, the rectum and some of the reproductive organs. The external iliac arteries continue into the thigh, where the name of these tubes is changed to **femoral** (fem'or-al). These vessels give off branches in the thigh and then become the **popliteal** (pop-lit'e-al) **arteries** which subdivide below the knee. The subdivisions include the **tibial arteries**, which extend into the ankle and foot.

OTHER PARTS AND SUBDIVISIONS OF SYSTEMIC ARTERIES

Just as the larger branches of a tree give off limbs of varying sizes, so the arterial tree has a multitude of subdivisions (Fig. 10.4). Hundreds of additional names might be included, but we shall mention only a few here. The hand receives blood that courses through the subclavian artery, which becomes the **axillary** (ak'si-lar-e) in the armpit. The longest part of this vessel is that in the arm proper. This portion is called the **brachial artery**. It subdivides into two branches near the elbow. These are the **radial artery**, which continues down the thumb side of the forearm and wrist, and the **ulnar artery**, which extends along the medial or little finger side into the hand. The common carotid artery

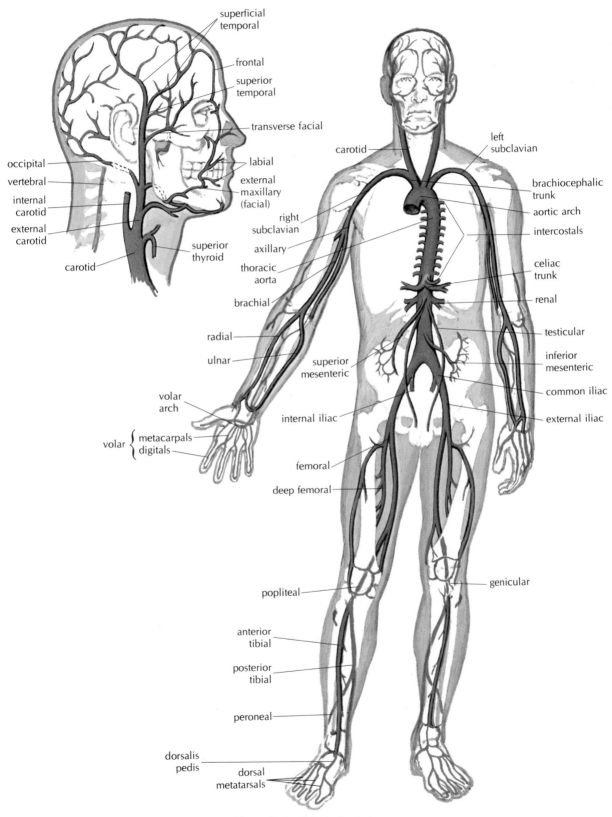

Figure 10.4. Principal arteries.

sends branches to the thyroid gland and other neck structures, and near the jaw it divides into an **external** and an **internal carotid**.

ANASTOMOSES

A communication between two arteries is called an **anastomosis** (ah-nas-to-mo′sis). By this means blood reaches vital organs via more than one route. Some examples of such unions of end arteries are:

1. The **circle of Willis**, which receives blood from the two internal carotid arteries as well as from the **basilar** (bas′i-lar) **artery**, which is formed by the union of two vertebral arteries. This arterial circle lies just under the center of the brain and sends branches to the **cerebrum**, the largest division of the brain, and to other parts of the brain.

2. The **volar** (vo′lar) **arch**, which is formed by the union of the radial and ulnar arteries in the hand. It sends branches to the hand and the fingers.

3. The **mesenteric arches**, which are made of communications between branches of the vessels that supply blood to the intestinal tract.

4. Arches which are formed by the union of the tibial arteries in the foot, and similar anastomoses which are found in various parts of the body.

Names of Systemic Veins

SUPERFICIAL VEINS

Whereas most arteries are located in protected and rather deep areas of the body, many veins are found near the surface. The most important of these superficial veins are in the two pairs of extremities (Fig. 10.5). These include:

1. The veins on the back of the hand and at the front of the elbow. Those at the elbow, incidentally, are often used for removing the blood for testing purposes, as well as for intravenous injections. The largest of this group of veins are the **cephalic** (se-fal′ik), **basilic** (bah-sil′ik) and the **median cubital** (ku′be-tal) **veins**.

2. The **saphenous** (sah-fe′nus) **veins** of the lower extremities, which are the longest veins of the body. The great saphenous vein begins in the foot and extends up the medial side of the leg, the knee and the thigh. It finally empties into the femoral vein near the groin.

DEEP VEINS

The deep veins tend to parallel arteries and usually have the same names as the corresponding arteries. Examples of these include the **femoral** and the **iliac** vessels of the lower part of the body and the **brachial**, the **axillary** and the **subclavian** vessels of the upper extremities. However, exceptions are found in the veins of the head and the neck. The **jugular** (jug′u-lar) **veins** drain the areas supplied by the carotid arteries. Two **brachiocephalic** (innominate) **veins** are formed, one on each side, by the union of the subclavian and the jugular veins. (Remember, there is but one brachiocephalic artery.)

SUPERIOR VENA CAVA

The veins of the head, the neck, the upper extremities and the chest all drain into the **superior vena cava** (ve′nah ka′vah), which goes to the heart. It is formed by the union of the right and the left brachiocephalic veins which drain the head, the neck and the upper extremities, while a special vein carries blood from the chest wall. The **azygos** (az′i-gos) **vein** drains the veins of the chest wall and empties into the superior vena cava just before the latter empties into the heart.

VENOUS SINUSES

The word "sinus" means "a space" or "a hollow." The sinusoids (the word means "like a sinus") found in the liver, the spleen, the thyroid gland and other structures are channels within the tissues of the organ. Larger channels which do not have the usual tubular structure of the veins also may drain deoxygenated blood. They are known as **venous sinuses** (see Fig. 10.6). An important example of a venous sinus is the **coronary sinus**, which receives most of the blood from the veins of the heart wall. It lies between the left atrium and left ventricle on the under (inferior) surface of the heart. It empties directly into the right atrium along with the two venae cavae.

Other important venous sinuses are lo-

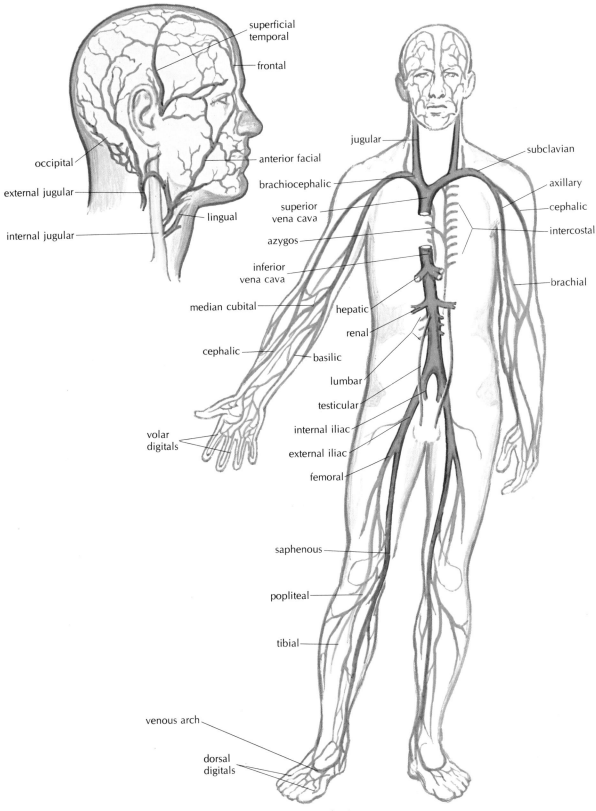

Figure 10.5. Principal veins.

cated inside the skull. They are the **cranial venous sinuses** which drain the veins that come from all over the brain. The largest of the cranial venous sinuses include:

1. The **cavernous sinuses**, situated behind the eyeball, which serve to drain the **ophthalmic veins** of the eye.

2. The **superior sagittal** (saj'i-tal) **sinus**, which is a single long space located in the midline above the brain and in the fissure between the two large hemispheres of the cerebrum. It ends in an enlargement called the **confluence** (kon'floo-ens) of sinuses.

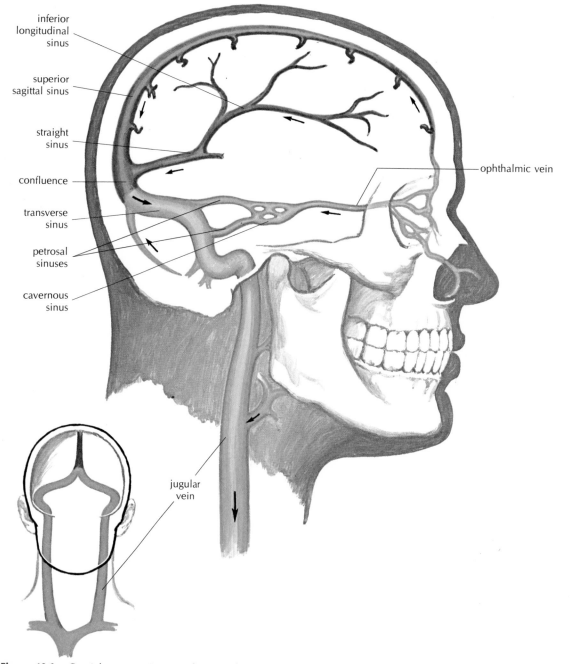

Figure 10.6. Cranial venous sinuses. The paired transverse sinuses, which carry blood from the brain into the jugular veins, are shown in light blue in the inset.

3. The two **transverse sinuses** which also are called the **lateral sinuses**. These sinuses are large spaces between the layers of the dura mater (a brain membrane) and extend toward each side after beginning at the center back in the region of the confluence of sinuses. As each sinus extends around the inside of the skull, it receives blood draining those parts not already drained by the superior sagittal and the other sinuses that join the back portions of the transverse sinuses. This means that nearly all the blood that comes from the veins of the brain eventually empties into one or the other of the lateral sinuses. On either side the sinus extends far enough forward to empty into an internal jugular vein, which then passes through a hole in the skull to continue downward in the neck. See inset of Fig. 10.6.

INFERIOR VENA CAVA

The **inferior vena cava** is much longer than the superior vena cava. The inferior vena cava returns the blood from the parts of the body below the diaphragm. It begins in the lower abdomen with the union of the two common iliac veins. It then ascends along the back wall of the abdomen, through a groove in the posterior part of the liver, through the diaphragm, and finally through the lower thorax to empty into the right atrium of the heart.

The drainage into the inferior vena cava is more complicated than drainage into the superior vena cava. We may divide the large veins below the diaphragm into two groups:

1. Those right and left veins that drain paired parts and organs. They include the **iliac veins** from near the groin; four pairs of **lumbar veins** from the dorsal part of the trunk and from the spinal cord; the veins from the testes of the male and the ovaries of the female called either the **testicular (spermatic) veins** or the **ovarian veins**; the **renal** and **suprarenal veins** from the kidneys and some glands near the kidneys; and finally the large **hepatic veins** from the liver. For the most part, these vessels empty directly into the inferior vena cava. The left testicular (spermatic) in the male and the left ovarian in the female empty into the left renal vein, which then takes this blood to the inferior vena cava; these veins thus constitute exceptions to the rule that the paired veins empty directly into the vena cava.

2. Unpaired veins which generally come from parts of the digestive tract (i.e., the bowel and the stomach) and empty into a special vein called the **portal tube**. This portal vein subdivides in the liver in order to permit blood from the spleen and the intestine to come into closer contact with the liver cells. Thus the complex functions of the liver may be performed (see Chapter 12).

THE PORTAL CIRCULATION

The unusual and exceptional feature of the portal vein is that after receiving the tributaries from the unpaired organs it enters the liver and subdivides into small vessels. All other veins continue to unite until they enter the tube (the vena cava) that carries the blood to the right atrium, or one of the four pulmonary veins that enter the left atrium. The largest tributary of the portal tube is the **superior mesenteric vein**. It is joined by the **splenic vein** just under the liver. Other tributaries of the portal circulation are the **gastric**, the **pancreatic** and the **inferior mesenteric veins**. The final subdivisions of the portal tube in the liver are spaces called **sinusoids** (si′nus-oids) which take the place of capillaries in the liver. The blood from the hepatic arterioles also enters these sinusoids. The blood in these spaces therefore is a combination of arterial and venous blood. It is collected finally by the hepatic veins, which empty into the inferior vena cava.

The portal circulation is interesting from another standpoint, because it is in this system that food products enter the blood circulation, to be carried eventually to the tissues. Food products are absorbed from the small intestine into the blood stream, and then travel to the liver through the portal vein. In the liver the food is altered, stored and released as needed into the main circulatory system (Fig. 10.7).

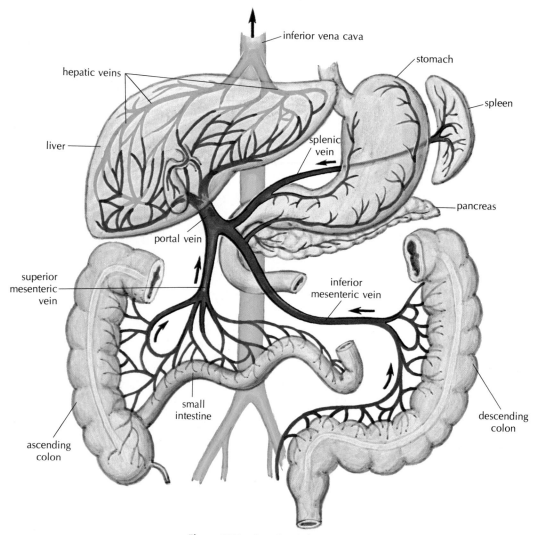

Figure 10.7. Portal circulation.

How Capillaries Work

We have spoken hitherto of the circulatory system as "bringing blood to the tissues." Actually, this is not entirely accurate, because blood does not touch the tissues directly (with the exception of the spleen).

If we think back to the discussion of the cells, it will be recalled that the cells are surrounded by a salty liquid called tissue fluid. We might think of the cells as a multitude of individual islands implanted in the middle of a lake. At the edge of this lake of tissue fluid runs a capillary, connected at one end to an artery and at the other to a vein (like the crossbar on the letter H). As the blood from the artery, charged with oxygen and food, passes through the capillary, those materials necessary for the life of the cells pass through the capillary walls into the lake of tissue fluid. From there they make their way to the cell islands. From the islands, in the opposite direction, come the waste products of cell metabolism. These pass into the capillary through its walls and proceed to the veins, whence they reach their organs of excretion (Fig. 10.8).

It must be remembered, of course, that the capillaries are tiny, hairlike vessels; that the letter H of which we spoke has not one

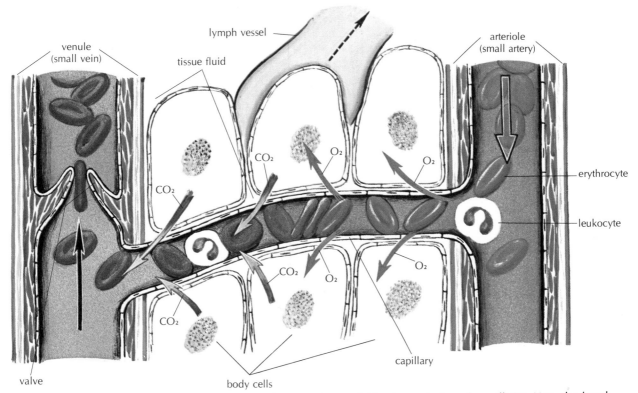

Figure 10.8. Diagram showing the connection between the small blood vessels through capillaries. Note the lymph capillary, a part of tissue drainage.

crossbar but millions of them. But the principle remains the same.

Pulse and Blood Pressure

MEANING OF THE PULSE

The ventricles pump blood into the arteries regularly about 70 to 80 times a minute. The force of ventricular contraction starts a wave of increased pressure which begins at the heart and travels along the arteries. This wave is called the **pulse**. It can be felt in the arteries that are relatively close to the surface, particularly if the vessel can be pressed down against a bone. At the wrist the radial artery passes over the bone on the thumb side of the forearm, and the pulse is most commonly obtained here. Other vessels sometimes used for obtaining the pulse include the carotid artery in the neck and the **dorsalis pedis** (dor-sa'lis pe'dis) of the top of the foot.

Normally the pulse rate is the same as the heart rate. Only if a heartbeat is abnormally weak may it be lost and thus not detected as a pulse motion. In checking the pulse of another person it is important to use the second or third fingers. If you use your thumb, you may find that you are getting your own pulse. A trained person can tell a great deal about the condition of the circulatory system and can gauge the strength as well as the regularity and the rate of the pulse.

Various factors may influence the pulse rate. We will enumerate just a few:

1. The pulse is somewhat faster in smaller people, and usually is slightly faster in women than in men.

2. In a newborn infant the rate may be from 120 to 140 beats per minute. As the child grows, the rate tends to become slower.

3. Muscular activity influences the pulse rate. During sleep the pulse may slow down to 60 a minute, while during strenuous exercise the rate may go up

to well over 100 a minute. If a person is in good condition, the pulse does not remain rapid despite a continuation of exercise.

4. Emotional disturbances may increase the pulse rate.

5. In many infections the pulse rate increases with the increase in temperature.

6. An excessive amount of secretion from the thyroid gland may cause a rapid pulse. The pulse rate may serve as a partial guide for the person who must take thyroid extract.

BLOOD PRESSURE AND ITS DETERMINATION

Since the pressure inside the blood vessels varies with the condition of the heart and the arteries as well as with other factors, the measurement of blood pressure together with careful interpretation may prove a valuable guide in the care and evaluation of a person's health. The pressure decreases as the blood flows from arteries into capillaries and finally into veins. Ordinarily, measurements are made of arterial pressure only. The instrument used is called a **sphygmomanometer** (sfig-mo-mah-nom'e-ter). The two measurements made are of:

1. The **systolic pressure**, which occurs during heart muscle contraction and averages around 120, expressed in millimeters of mercury.

2. The **diastolic pressure**, which occurs during relaxation of the heart muscle and averages around 80 millimeters of mercury.

The sphygmomanometer is essentially a graduated column of mercury connected to an inflatable cuff. The cuff is wrapped around the patient's upper arm and is inflated with air until the brachial artery is compressed and the blood flow cut off. Then, listening with a stethoscope, the doctor or nurse slowly lets air out of the cuff until the first pulsations are heard. At this point the pressure in the cuff is equal to the systolic pressure; and this pressure is read off the mercury column. Then, more air is let out until another characteristic sound indicates the point at which the diastolic pressure is to be read off. Considerable practice is required to insure an accurate reading.

VARIATIONS IN BLOOD PRESSURE

Normal systolic presure may vary from 90 to 140 mm. of mercury, while diastolic pressure ranges between 50 and 90 mm. Lower than normal blood pressure is called **hypotension** (hi-po-ten'shun). Many apparently healthy persons have systolic blood pressures below 110. The sudden lowering of blood pressure is an important symptom of shock. **Hypertension** (hi-per-ten'shun), which is high blood pressure, has received a great deal of attention. It often occurs temporarily as a result of excitement or exertion. Although stress has been placed on the systolic blood pressure, in many cases the diastolic pressure is even more important. The condition of the small arteries may have more effect on the diastolic pressure.

Some of the factors that are important in the maintenance of normal blood pressure include the efficiency of the heart muscle, the resistance in the blood vessels, the blood volume, and the condition of the artery walls.

Summary

1. **Functional classification of blood vessels.**
   A. Arteries carry blood from the heart to other parts of the body. Smallest subdivisions are arterioles.
   B. Veins drain tissues and return blood to the heart. Smallest subdivisions are venules.
   C. Capillaries allow exchanges between blood and body cells. These take place through the tissue fluid.

2. **Two circuits.**
   A. Pulmonary vessels: connected with the lungs.
   B. Systemic vessels: form a network in all other parts of the body.

3. **Structure of blood vessels.**
   A. Arteries have thick walls in 3 layers: endothelium, involuntary muscle and connective tissue.

B. Capillaries are a single cell layer thick, a continuation of the endothelial lining of larger vessels.

C. Veins have 3-layer walls but are more easily collapsed than arteries because of their thinner walls, having less muscle and elastic connective tissue. Most have one-way valves.

4. **Names of systemic arteries.**
   A. Parts of the aorta are the ascending aorta, aortic arch, thoracic aorta and abdominal aorta.
   B. Branches of the ascending aorta are the 2 coronary arteries that supply the heart muscle.
   C. Three branches of the aortic arch.
      (1) Brachiocephalic trunk (formerly the innominate artery).
      (2) Left common carotid artery.
      (3) Left subclavian artery.
   D. Branches of the thoracic aorta.
      (1) Esophageal artery.
      (2) Bronchial arteries.
      (3) Intercostal arteries: 9 or 10 pairs.
   E. Branches of the abdominal aorta.
      (1) Unpaired branches to the viscera.
         (a) Celiac artery to stomach, spleen and liver.
         (b) Superior mesenteric to small intestine and first half of large intestine.
         (c) Inferior mesenteric artery to last half of large intestine.
      (2) Paired branches.
         (a) Phrenic arteries to the diaphragm.
         (b) Suprarenal arteries to the adrenal (suprarenal) glands.
         (c) Renal arteries to the kidneys.
         (d) Ovarian arteries or testicular (spermatic) arteries to the sex glands.
         (e) Lumbar arteries (4 pairs) to muscles of abdominal wall.
      (3) Iliac arteries.
         (a) Internal (hypogastric) arteries to pelvic organs.
         (b) External (femoral) arteries to thigh, which continue as the popliteal arteries. These subdivide below knee; latter group includes tibial arteries.
   F. Other arteries and structures.
      (1) Parts of subclavian artery include axillary and brachial arteries.
         (2) Anastomoses are unions of end arteries.
            (a) Circle of Willis to the brain.
            (b) Volar arch in the hand to supply fingers.
            (c) Mesenteric arches to the bowel.
            (d) Arches formed by the tibial arteries.

5. **Names of systemic veins.**
   A. Superficial veins near the surface in extremities.
      (1) Cephalic, basilic, and the median cubital veins, all in the upper extremities.
      (2) Saphenous veins in lower extremities.
   B. Deep veins accompany or parallel arteries and have the same names as the arteries.
      (1) Examples: femoral, iliac, brachial, axillary, subclavian.
      (2) Jugular veins from head and neck are important exceptions to the naming rule.
      (3) Superior vena cava receives all tributaries from above the diaphragm.
      (4) Inferior vena cava receives paired vessels from below diaphragm.
   C. Portal circulation.
      (1) Portal tube receives unpaired vessels from spleen, bowel and stomach.
      (2) Portal tube subdivides in the liver until sinusoids are formed.
      (3) Blood of sinusoids collected by the hepatic veins which empty into the inferior vena cava.

6. **Venous sinuses.**
   A. Coronary sinus.
   B. Cranial sinuses. Largest are 2 cavernous, a single superior sagittal and 2 transverse (lateral) sinuses.

7. **Pulse and blood pressure.**
   A. Pulse is a wave in the arteries due to contraction of the heart muscle.
   B. Pulse varies with the size, sex, age and activity of the healthy person.
   C. Pulse rate may be increased by emotional disturbances, fever or excessive thyroid secretion.
   D. Blood pressure systolic during heart contraction, diastolic during relaxation of ventricles.
   E. Stethoscope and sphygmomanometer used for obtaining readings.
   F. Normal blood pressure influenced by efficiency of the heart muscle, resistance in the blood vessels, blood volume and condition of the arterial walls.

Questions and Problems

1. Name the 3 main groups of blood vessels and describe their functions. How has function affected structure?

2. Trace a drop of blood through the shortest possible route from the capillaries of the foot to the capillaries of the head.

3. What are the names and functions of some cranial venous sinuses? Where is the coronary venous sinus and what does it do?

4. What large vessels drain the blood low in oxygen from most of the body into the right atrium? What vessels carry blood high in oxygen into the left atrium?

5. How does the portal vein differ from other veins? Why?

6. What is meant by pulse? Where is it most often obtained? If a large part of the body were burned, leaving only the lower extremities accessible for obtaining the pulse, what vessel would you try to use?

7. What are some factors that cause an increase in the pulse rate?

8. What is the usual effect of emotional disturbances or physical activity on the pulse? How long should these effects disturb the healthy person as he continues to exercise, for example?

9. What instrument is used for obtaining the blood pressure? What are the two values usually obtained called and what is the significance of each?

10. Describe some of the factors that influence normal blood pressure.

The Lymphatic System and Lymphoid Tissue

The Lymphatic System

It may be recalled, from the section called "How Capillaries Work" in the preceding chapter, that exchanges are carried on between the islandlike groups of cells and the capillaries across the lake of tissue fluid. The food and oxygen that pass through the capillary walls into the tissue fluid are dissolved in water and leave the blood plasma as a solution; consequently, the volume of tissue fluid is constantly being added to, and some provision must be made for draining off this excess fluid in order to keep the level constant.

One drainage pathway for the tissue fluid is the **lymphatic system**. Besides the capillary carrying blood, there is another pipeline at the shore of the lake of tissue fluid which we had not hitherto noticed. It is called a **lymphatic capillary**, and is nothing more than a drainpipe (see Fig. 10.8). As the cells send waste products across the lake to be taken away and disposed of, some of the waste products pass through the walls of the blood (vascular) capillary and go directly to the veins. However, certain other waste products pass into the lymphatic capillary along with the excess tissue fluid. This fluid, as it passes into the lymphatic capillary, is now called **lymph**. The lymphatic capillaries join to form the larger **lymphatic vessels**, and these vessels (which we shall have a closer look at in a moment) eventually empty into the veins. However, before the lymph reaches the veins, it is passed through a series of filters called **lymph nodes**, where certain solid particles are taken out. Thus the lymph nodes may be compared in one way with the oil filter in an automobile.

Let us now have a closer look at the lymphatic capillaries, vessels and nodes.

LYMPHATIC CAPILLARIES

The lymphatic capillaries resemble the blood capillaries in that they are made of one layer of flattened cells to allow for easy passage of soluble materials and water through

131

them. Unlike the capillaries of the blood stream, the lymphatic capillaries begin blindly; that is, they do not serve to bridge two larger vessels. Instead, one end simply lies within the lake of tissue fluid, while the other communicates with the larger lymphatic vessel.

In the intestine are some specialized lymphatic capillaries called **lacteals** (lak'te-als) which act as one pathway for the transfer of fats from digested food to the blood stream. This process is covered in the chapter dealing with the digestive system. A lacteal is pictured in Fig. 12.4.

LYMPHATIC VESSELS AND RIGHT LYMPHATIC DUCT

The lymphatic vessels are thin-walled and delicate, and have a beaded appearance because of indentations at the regions at which valves are located. These valves prevent backflow in the same way as those which are found in some veins. Exercise and changes in the position of parts of the body help in maintaining the flow of lymph; there is no "lymph heart."

Lymphatic vessels include **superficial** and **deep** sets. The surface lymphatics are immediately below the skin, often continuing near the superficial veins. The deep vessels are usually larger, and accompany the deep veins. All the lymphatic vessels form networks, and at certain points they carry lymph into the regional lymph nodes (i.e., the nodes which "service" a particular area). Lymphatic vessels are named according to their location. For example, those on the lateral or thumb side of the forearm are called **radial lymphatic vessels**, while those in the medial part of the forearm are called the **ulnar lymphatic vessels**. Nearly all of the lymph from the upper extremity and from the breast is carried into the **axillary lymph nodes** (i.e., in the axilla, or armpit). From the axillary lymph nodes, lymphatic vessels extend to one of the two large drainage ducts which finally empty into the blood stream.

The **right lymphatic duct** is a short tube about ½ inch long (1.25 cm.) which receives the lymph that comes from only the right side of the head, the neck and the thorax, and also from the right upper extremity. It empties into the right subclavian vein. Its opening into this vein is guarded by two pocketlike semilunar valves to prevent blood from entering the duct. The rest of the body is drained by the **thoracic duct**.

THE THORACIC DUCT

All parts of the body except those above the diaphragm on the right side are drained by the thoracic duct, which is the larger of the two main lymph channels. This tube is about 16 inches long (40 cm.). It begins at the back of the abdomen and below the attachment of the diaphragm. The first part of this tube is enlarged to form a cistern or temporary storage pouch called the **cisterna chyli** (sis-ter'nah ki'li). **Chyle** (kile) is the milky-appearing fluid formed by the combination of fat globules and lymph, which comes from the intestinal lacteals. Chyle passes through the intestinal lymphatic vessels and the lymph nodes of the mesentery, finally entering the cisterna chyli. In addition to chyle, all the lymph from below the diaphragm empties into the cisterna chyli by way of the various clusters of lymph nodes and then is carried by the thoracic duct into the blood stream.

The thoracic duct extends upward through the diaphragm and along the back wall of the thorax up into the root of the neck on the left side. Here it receives the left jugular lymphatic vessels from the head and the neck, the left subclavian vessels from the left upper extremity, and other lymphatic vessels from the thorax and its parts. In addition to the valves along the duct, there are two at its entrance into the left subclavian vein. Thus lymph and fat enter the blood stream to be distributed to all parts of the body (see Fig. 11.1).

Structures of Lymphoid Tissue

The foregoing section is a brief survey of the system of lymph vessels and lymph conduction. The lymph nodes, or filters, have been mentioned repeatedly; but detailed discussion of them has been withheld until now. There has been a reason for this. The lymph nodes are made of a specialized tissue called **lymphoid** (lim'foid) **tissue**. A number of other organs also are made of lymphoid tissue, but some of them have nothing directly to do with the system of lymph conduction itself. Therefore, it seems advisable to look at

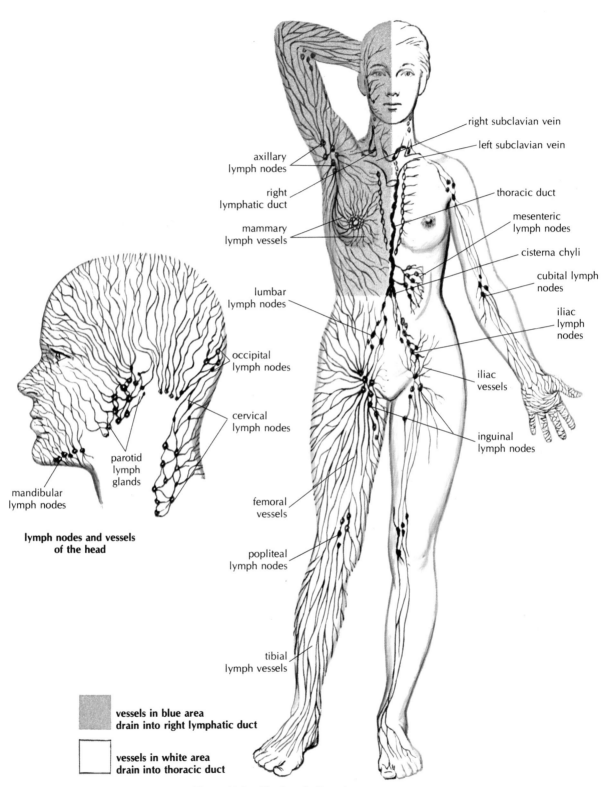

right subclavian vein

left subclavian vein

axillary
lymph nodes

right
lymphatic duct

thoracic duct

mesenteric
lymph nodes

mammary
lymph vessels

cisterna chyli

cubital lymph
nodes

lumbar
lymph nodes

iliac
lymph
nodes

occipital
lymph nodes

iliac
vessels

cervical
lymph nodes

inguinal
lymph nodes

parotid
lymph
glands

mandibular
lymph nodes

femoral
vessels

popliteal
lymph nodes

**lymph nodes and vessels
of the head**

tibial
lymph vessels

vessels in blue area
drain into right lymphatic duct

vessels in white area
drain into thoracic duct

Figure 11.1. The lymphatic system.

these lymphoid tissue structures as a group, taking the lymph nodes first so that the lymph conduction system will have received some consideration before the other "filter masses" are studied.

Before looking at some typical organs of lymphoid tissue, let us first consider some properties of this kind of tissue and see what characteristics these organs have in common.

FUNCTIONS OF LYMPHOID TISSUE

Some of the general functions of lymphoid tissue include:

1. Removal of impurities such as carbon particles, cancer cells, pathogenic organisms and dead blood cells.
2. Manufacture of lymphocytes, which make up 20 to 25 percent of the white blood cells.
3. Production of **antibodies**, which are chemical substances that aid in combatting infection.

LYMPH NODES

The lymph nodes, as we have seen, are designed to filter the lymph once it is drained from the tissues. The lymph nodes are small rounded masses varying from pinhead size to as much as an inch in length (1 to 25 mm.). Each node has a fibrous connective tissue capsule from which partitions extend into the substance of the organ. Inside the node are masses of lymphatic tissue, with spaces set aside for the production of lymphocytes. At various points in the surface of the node lymphatic vessels pierce the capsule in order to carry lymph into the spaces inside the pulplike nodal tissue. An indented area called the **hilum** (hi'lum) serves as the exit for lymph vessels carrying lymph out of the node. At this region other structures, including blood vessels and nerves, connect with the organ.

Lymph nodes seldom are isolated. As a rule they are massed together in groups, the number in each group varying from two or three up to well over 100. Some of these groups are placed deeply, while others are superficial. Those of the most practical importance include:

1. **Cervical nodes**, located in the neck. They are divided into deep and superficial groups, which drain various parts of the head and the neck. They often become enlarged during upper respiratory infections as well as in certain chronic disorders.
2. **Axillary nodes**, located in the armpits. They become enlarged from infections of the upper extremities and the breasts. Cancer cells from the breasts (mammary glands) often invade these lymph nodes, and so must be removed at the time the breast is removed in order to prevent the further spread of cancer.
3. **Tracheobronchial** (tra-ke-o-brong'ke-al) **nodes**, found near the trachea and around the larger bronchial tubes. In city dwellers these nodes become so filled with carbon particles that they are solid black masses resembling pieces of coal.
4. **Mesenteric** (mes-en-ter'ik) **nodes**, found between the two layers of peritoneum that form the mesentery. There are some 100 to 150 of these nodes.
5. **Inguinal nodes**, located in the groin region. They receive lymph drainage from the lower extremities and from the external genital organs. When they become enlarged, they are often referred to as **buboes** (bu'boes). The name for **bubonic plague**, which killed so many people during the Middle Ages, arose from the fact that the bacteria caused enlargement of various lymph nodes, especially the inguinal ones.

This discussion of the lymph nodes completes our short outline of lymph circulation. The following organs of lymphoid tissue perform somewhat different functions, particularly with respect to the substances which they filter.

THE TONSILS

There are other masses of lymphoid tissue which are designed to filter not lymph, but tissue fluid. These masses are found beneath certain areas of moist epithelium which are exposed to the outside, and hence to contamination. Such areas include parts of the digestive, the urinary and the respiratory tracts. With this last-named system are associated those well-known masses of lymphoid tissue called the **tonsils.**

The different tonsils include:

1. The **palatine** (pal'ah-tin) **tonsils**, which are oval bodies located at each side of the soft palate. These are most commonly known as the tonsils.
2. The **pharyngeal** (fah-rin'je-al) **tonsil**, which, when enlarged, is commonly referred to as **adenoids**. It is located behind the nose on the back wall of the upper pharynx.
3. The **lingual** (ling'gwal) **tonsils**, which are little mounds of lymphoid tissue at the back of the tongue.

Any or all of these tonsils may become so loaded with bacteria that the pathogens come to have the upper hand; removal then is certainly advisable. A slight enlargement of any of them will not call for surgery. All lymphoid tissue masses tend to be larger in childhood, so that consideration must be made of the patient's age in determining whether or not these masses are enlarged abnormally.

THE THYMUS

Because of its appearance under the microscope, the **thymus** (thi'mus) has been considered a part of the lymphoid system. Recent studies point to its having a much more basic function than was originally thought. It now seems apparent that the thymus plays a key role in the formation of antibodies in the first few weeks of life and in the development of immunity. It manufactures lymphocytes and is essential to fetal growth. It seems that the factors which stimulate the formation of lymphocytes come from the thymus itself. Removal causes a decrease in the production of these cells, and a decrease in the size of the spleen and of lymph nodes throughout the body. The thymus is most active during early life. After puberty, the tissue undergoes changes and is replaced by adipose tissue (see Chap. 15).

THE SPLEEN

The spleen is an organ which contains lymphoid tissue designed to filter blood. It is located in the upper left hypochondriac region of the abdomen and normally is protected by the lower part of the rib cage because it is high up under the dome of the diaphragm. The spleen is rather soft and of a purplish color. It is a somewhat flattened organ about 5 or 6 inches long (12.5 to 15 cm.) and 2 or 3 inches wide (5 to 7.5 cm.). The capsule of the spleen, as well as its framework, is more elastic than that of the lymph nodes. It contains involuntary muscle which makes it possible for the splenic capsule to contract, and also makes it able to withstand some swelling.

The spleen has an unusually large blood supply, considering its size. The organ is filled with a soft pulp, one of the functions of which is to filter out the worn-out red blood cells. Also generated in the spleen are numbers of white blood cells of the type that are capable of **phagocytosis**, which, as mentioned in Chapter 4, is the process of engulfing bacteria and other foreign cells. Prominent structures inside the spleen are round masses of lymphoid tissue, and it is because of these that the spleen often is classified with the other organs made of lymphoid tissue (although some other category might be better because of the other specialized functions of the spleen).

Some of the functions of the spleen are:

1. To destroy used-up **erythrocytes** (red blood cells) and to return some of the products of this decomposition to the liver via the portal tube, so that the liver can use them in the manufacture of bile (an important secretion in the process of fat digestion).
2. To serve as a reservoir for blood which can be supplied to the blood stream by contraction of the capsule in case of a hemorrhage or some other emergency.
3. To produce lymphocytes and another type of white cell (monocytes).
4. To produce red cells before birth (this function is abandoned afterward).
5. To aid in removing all sorts of foreign and undesirable matter from the tissues, as do parts of the lymph nodes, the bone marrow, and the liver also. All of these organs possess cells called **macrophages** (mak'ro-faj-es), which absorb and destroy this foreign matter.
6. To produce antibodies which lend immunity to certain diseases.
7. To remove iron from hemoglobin and to direct it for reuse by the bone marrow in manufacturing new red blood cells.

135

A final comment on the characteristics of lymphoid tissue: nature has furnished the human organism with thousands of these masses, many more than are required to combat infection; and so there is a large reserve. Tonsils may become infected and removal of them for that reason is common. The spleen may need to be removed in certain disorders. Lymph nodes may be invaded by cancer and should be completely extirpated (removed). No one group or mass of these tissues is necessary for life.

Summary

1. **Lymph.**
   A. Fluid inside of lymphatic vessels.
   B. Formed from tissue fluid.
   C. Takes away some cellular waste products.
   D. Goes through capillaries and vessels, is filtered in nodes, returns to blood.
2. **Lymphatic conduction.**
   A. Lymphatic capillaries drain lymph directly from tissues (some capillaries in intestines are lacteals).
   B. Lymphatic vessels receive lymph from capillaries. Have valves; lymph flow caused by muscular action.
   C. Lymphatic vessels lead through lymph nodes (filters) and then to 2 main ducts: right lymphatic, thoracic.
   D. Right lymphatic duct drains right side of the body above diaphragm. Empties into right subclavian vein.
   E. Thoracic duct drains rest of body. Has cisterna chyli, an enlargement at its beginning. Empties into left subclavian vein.
3. **Lymphoid tissue.**
   A. Found in other structures than those in lymph conduction system.
   B. Functions: removes bacteria and other foreign particles, manufactures lymphocytes, produces antibodies.
   C. Individual masses of tissue expendable.
4. **Lymph nodes.**
   A. Act as lymph filters.
   B. Main groups: cervical, axillary, tracheobronchial, mesenteric, inguinal.
5. **Tonsils.**
   A. Are but 1 example of a group of lymphoid masses located beneath moist exposed epithelium.
   B. Forms: palatine, pharyngeal, lingual.
6. **Thymus.**
   A. Most active during early life.
   B. Plays role in development of immunity.
   C. Manufactures lymphocytes.
7. **Spleen.**
   A. Soft flattened organ.
   B. Functions: filters blood, is reservoir for blood, produces erythrocytes in embryo, produces lymphocytes and monocytes, destroys bacteria and other foreign matter, produces antibodies, removes iron from hemoglobin and directs it for reuse in erythrocyte formation.

Questions and Problems

1. What is lymph? Name some of its purposes.
2. Briefly describe the system of lymph circulation.
3. Describe the lymphatic vessels with respect to design, appearance, and depth of location.
4. Name the 2 main lymphatic ducts. What part of the body does each drain, and into what blood vessel does each empty?
5. What is the cisterna chyli and what are its purposes?
6. Name 3 characteristics of lymphoid tissue.
7. Describe the structure of a typical lymph node.
8. What are the neck nodes called and what are some of the causes of enlargement of these lymph nodes?
9. What parts of the body are drained by vessels entering the axillary nodes and what conditions cause enlargement of these nodes?
10. What parts of the body are drained by lymphatics that pass through the inguinal lymph nodes? What is the relationship of bubonic plague to these nodes?
11. What are the different tonsils called and where are they located? What is the purpose of these and related structures?
12. What is the function of the thymus?
13. Give the location of the spleen and name several of its functions.

The Digestive System

What the Digestive System Does

In this chapter we shall study the mechanism by which the food that we eat manages to reach the cells of every part of the body. This process is not so simple as it might seem. A solitary cell would be baffled if a fragment of food, in the state that is familiar to us, appeared across the lake of tissue fluid and sought admission. Food must be converted to a state in which it is capable of being taken into the cells by way of the blood plasma. This conversion process is known as **digestion**. Once the food is digested, however, it must be transferred to the blood stream; and the process by which this transfer occurs is called **absorption**. Digestion and absorption are the two chief functions of the digestive system (Fig. 12.1).

For purposes of study, the digestive system may be divided into two groups of organs:

1. The **alimentary canal**, which is a continuous passageway beginning at the mouth, where food is taken in, and terminating at the anus, where the solid waste products of digestion are expelled from the body.
2. The **accessory organs** which, while vitally necessary for the digestive process, do not happen to be part of the alimentary canal.

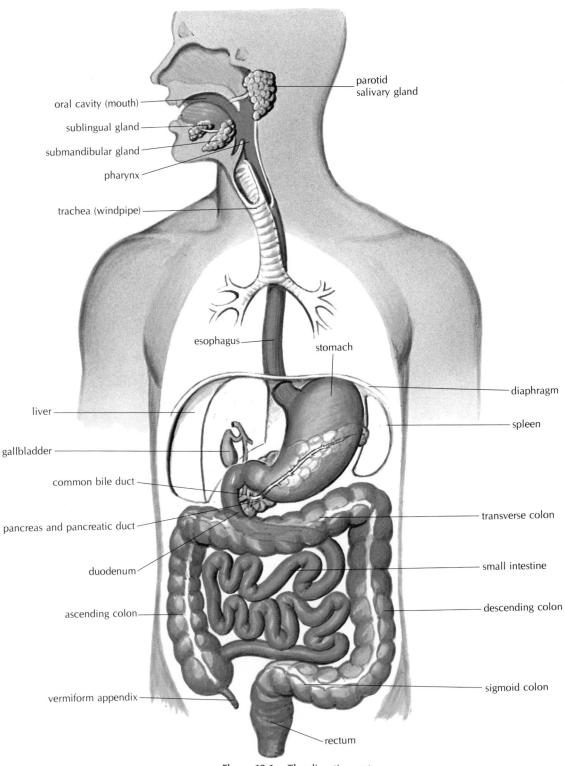

oral cavity (mouth)

parotid salivary gland

sublingual gland

submandibular gland

pharynx

trachea (windpipe)

esophagus

stomach

diaphragm

liver

spleen

gallbladder

common bile duct

transverse colon

pancreas and pancreatic duct

duodenum

small intestine

ascending colon

descending colon

sigmoid colon

vermiform appendix

rectum

Figure 12.1. The digestive system.

The Alimentary Canal

The **alimentary** (al-e-men'tar-e) **canal** is a muscular digestive tube extending through the body. It is composed of several parts: the **mouth**, the **pharynx**, the **esophagus**, the **stomach**, the **small intestine** and the **large intestine**, all of which will be defined and described as we encounter them. In this chapter there will be a separate section devoted to the accessory organs, although we shall be familiar with at least the main functions of all of them by the time we have finished our survey of the alimentary canal.

The word "aliment" comes from a Latin word which means "food" or "nutrients." Those foods which undergo changes and are absorbed into the blood leave the tube from the region of the small intestine. Indigestible substances such as coins, safety pins, and the cellulose in food pass the whole distance through the alimentary canal and are expelled from the body.

THE MOUTH, OR ORAL CAVITY

The mouth is also called the oral cavity. A digestible lump of sugar as well as an indigestible coin would begin the tour of the alimentary canal in this cavern. The oral cavity has three purposes:
1. To receive food.
2. To prepare food initially for the digestive process.
3. To aid in the accomplishment of speech.

Into this space there projects a muscular organ called the **tongue**. The tongue is an aid to chewing and swallowing, and in addition is one of the principal organs of speech. (The Latin word "lingua," meaning "tongue," gave rise to our word "language.") The tongue has on its surface a number of special organs called **taste buds**, by means of which we can differentiate taste sensations (bitter, sweet, sour or salty).

Within this cavity are also a number of stonelike structures, the teeth. There are 20 of these stony structures in a child between two and six years of age. The adult with the good fortune to have a complete set would have 32 teeth. The cutting teeth, or **incisors** (in-si'zers), occupy the front part of the buccal cavity, while the larger grinding teeth, called the **molars**, are in the back portion of the oral cavity (see Fig. 12.2).

Deciduous, or baby, teeth

The first eight **deciduous** (de-sid'u-us) **teeth** to make their appearance through the gums are the incisors. Later the **canines** (eye-teeth) and molars appear. Usually the 20 baby teeth have all made their appearance by the time the infant has passed his second birthday. During this time the permanent teeth continue development within the jawbones. The first permanent tooth to make its appearance is the very important six-year molar. This permanent tooth comes in before the baby incisors are lost, and many people do not realize that a key permanent tooth has appeared. Decay and infection of the adjacent deciduous molars may spread to and involve the new permanent molar tooth. Deciduous teeth need proper care in order to help preserve the six-year molars and other permanent teeth.

Permanent teeth

Although the buds for the second set of teeth are present at birth, the first permanent tooth does not usually make its appearance in the mouth until the child is about six years old. At that time the first molar, the keystone for the future grinding surfaces, appears in the space now present behind the baby molars. As the child grows, the jawbones grow also; therefore there is space for more teeth than are in the first set. After the first permanent molar has made its appearance, the baby incisors loosen and are replaced by **permanent incisors**. Then the baby canines (cuspids) are replaced by **permanent canines**, and finally the baby molars are replaced by the bicuspids (premolars) of the permanent set. Now the larger jawbones are ready for the appearance of the 12-year or second permanent molar teeth. Somewhat later, the third molars, or so-called **wisdom teeth**, appear. In some cases the jaw is not large enough, or there are other abnormalities, so that these teeth may have to be removed early in life.

Diseases of the mouth and the teeth

Infection of the gum is called **gingivitis** (jin-je-vi'tis), while infection of the rest of the mucous lining of the mouth is called

stomatitis (sto-mah-ti'tis). Stomatitis has recently become a problem for people who use antibiotic types of lozenges. These medicated wafers may encourage fungous infections of the mouth and the tongue. **Vincent's angina** is a kind of gingivitis, causing redness and ulceration of the mucous membrane of the mouth and gums. It is contagious and is caused by a spirochete. **Pyorrhea** is an inflammation involving the tooth socket or **alveolus** (al-ve'o-lus). It is accompanied by discharge of pus, so the name is really **pyorrhea alveolaris** (pi-o-re'ah al-ve-o-la'ris). "Pyo" means "pus," "rrhea" means "flow" or "discharge," while "alveolaris" refers to a canal or socket.

Tooth decay or dental **caries** (ka're-ez), which means "rottenness," has a number of causes. For Americans, it is a particularly expensive disease. Primitive peoples and others who do not indulge in the elegant foods so characteristic of the American diet have much less tooth decay. In addition to diet, such factors as heredity, mechanical problems and endocrine disorders are believed to play a part. Since a baby's teeth begin development before birth, the diet of the mother during pregnancy also is most important in forestalling tooth decay.

The salivary glands

Another contribution to the digestive mechanism furnished by the oral cavity is the production of **saliva**. The purpose of saliva is to dissolve the food and to facilitate the processes of chewing (**mastication**) and swallowing (**deglutition**). Saliva also coats the food with mucus, allowing it to "go down" more easily. The chemical function of saliva will be discussed later in this chapter.

Saliva is manufactured by three pairs of glands, the first of the accessory organs:
1. The **parotid** (pah-rot'id) glands, the largest of the group.
2. The **submandibular** (sub-man-dib'u-lar), or **submaxillary** (sub-mak'si-ler-e), glands, located near the body of the lower jaw.
3. The **sublingual** glands, under the tongue.

The parotid salivary glands, located near the ear, are infected in the contagious disease commonly called **mumps**. The infecting agent is a virus. The patient should remain quiet in the hope of preventing the spread of the infection to the sex glands, the pancreas and the other salivary glands. As is true of many contagious diseases mumps is now preventable by the use of a vaccine.

THE WALLS OF THE ALIMENTARY CANAL

Beyond the oral cavity, the walls of the parts of the alimentary canal from the swallowing tubes to the anus are all of a more or less uniform design. First of all, the canal is lined with mucous membrane. Beneath the mucosa is a layer of "packing" tissue (connective) containing blood vessels and nerves. Next come layers of involuntary muscle tissue with a most interesting function. When food reaches the swallowing tubes, this muscle tissue is stimulated to produce a rhythmic wavelike motion known as **peristalsis** (per-e-stal'sis), as a result of which the food is transported the entire length of the alimentary canal and mixed with digestive juices en route.

The final layer of the alimentary canal is of fibrous connective tissue—except for those parts that extend into the abdominal cavity, which have an additional layer called peritoneum—and will be discussed later in this chapter.

THE SWALLOWING TUBES AND THEIR ACCESSORIES

The **pharynx** is often referred to as the **throat**. The digestible food and the indigestible coin are both pushed by the tongue into the pharynx. The tongue and the walls of the pharynx are made of voluntary muscle with a lining of mucosa. The tonsils may be seen at either side of the pharynx. The **soft palate** forms the back of the roof of the oral cavity. From it hangs a soft, fleshy, V-shaped mass called the **uvula** (u'vu-lah) (see Fig. 12.2). This muscular tissue guards the opening from the nasal cavity and the upper pharynx, preventing foods and liquids from entering the nasal cavities. During the process of swallowing, the muscles of the pharynx contract and so constrict the space. At this time the openings into the air spaces both above and below the mouth are closed off, by the soft palate above and by the **epiglottis**, shaped like a small lid, below.

The **esophagus** (e-sof'ah-gus), or **gullet**, receives the contents of the contracting pharynx

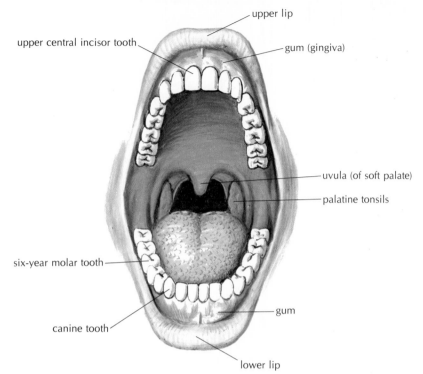

upper lip

upper central incisor tooth

gum (gingiva)

uvula (of soft palate)

palatine tonsils

six-year molar tooth

gum

canine tooth

lower lip

Figure 12.2. The mouth, showing the teeth and the tonsils.

and forces them on by peristalsis. The esophagus, about 9 inches (22.5 cm.) long, extends through the neck and the chest (thorax). Finally the esophagus reaches the abdominal cavity after extending through the diaphragm. There it empties into a saclike structure called the **stomach.**

THE STOMACH

The stomach is actually an enlarged (dilated) section of the alimentary tube (Fig. 12.3). It is shaped somewhat like a gourd, and both ends of it are guarded by valves which normally permit the passage of substances in only one direction. The first of these is the **cardiac valve**, located between the esophagus and the stomach. We frequently are aware of the existence of this valve; sometimes it does not relax as it should, and then there is a feeling of having a place one can't swallow past. At the distal or far end of the stomach, connecting it with the small intestine, is the other valve called the **pyloric sphincter** (pilor'ik sfingk'ter). This valve is especially important in that it determines the length of

time in which the food remains in the stomach.

The stomach, so often abused and misunderstood, is a combination storage pouch and churn. If the stomach is empty, there will be many folds in the lining. These folds are called **rugae** (ru'ge), and they disappear as the stomach fills (it may be stretched so that it holds a half gallon of food and liquid). When the stomach is filled, the pyloric sphincter closes and retains the contents until the food has been mixed with certain digestive juices collectively called **gastric juice**. These juices are secretions given out by tiny glands in the stomach wall. The mixture of gastric juice and food is known as **chyme** (kime).

The gastric juice itself has two main components: **hydrochloric acid** and **enzymes**.

Stomach acid

The hydrochloric acid in the stomach juice has three important functions:

1. It softens the connective tissues in meat.

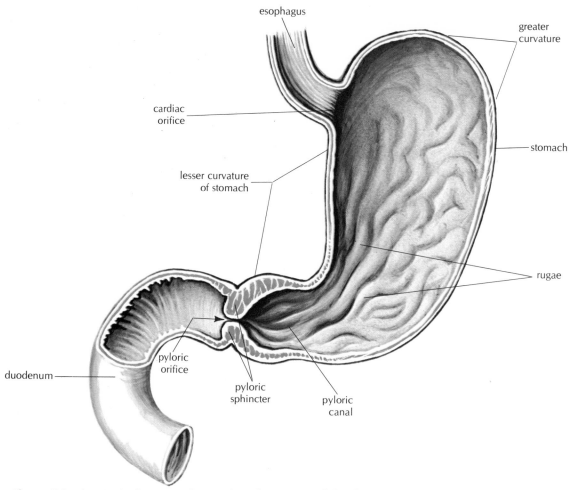

esophagus

greater
curvature

cardiac
orifice

stomach

lesser curvature
of stomach

rugae

duodenum

pyloric
orifice

pyloric
sphincter

pyloric
canal

Figure 12.3. Longitudinal section of stomach and a portion of the duodenum, showing interior including the valves (sphincters).

2. It kills bacteria and thus destroys many potential disease-producing agents.

3. It activates at least one of the stomach enzymes, which are chemicals that convert food into soluble and absorbable substances (see Chapter 2).

An abnormally low production of stomach acid may cause digestive disturbances which are greatly aggravated by taking soda or other alkaline substances contained in many patent medicines purporting to relieve indigestion. Such substances neutralize the valuable normal functions of the stomach acid, and in many cases grave harm is done by such self-medication. Occasionally, hydrochloric acid is produced in excess, but this condition is rather uncommon and can be diagnosed easily by an analysis of the stom-ach contents. What the layman thinks is excess acidity (**hyperacidity**) may really be underacidity (**hypoacidity**) and may indicate serious disease, possibly even cancer.

More about the pyloric valve

The pyloric valve (or sphincter) is a ring-like muscle surrounding the end of the stomach. Normally the stomach contents escape through this valve in about two to six hours. The exit valve from the stomach opens into the **small intestine.**

THE SMALL INTESTINE

The small intestine is by far the longest part of the alimentary canal. More appropriately, it might be called the long intestine, because it is about four or five times as long as the large intestine. However, it is known as the small intestine because of its smaller diameter. The small intestine averages about 20 feet in length as compared with some four or five feet for the large intestine. In addition, the mucosa which lines the small intestine is greatly increased in area by tiny, fingerlike projections called **villi** (vil'li) (see Fig. 12.4). The first 10 or 12 inches (25 to 27 cm.) of the small intestine is called the **duodenum**. In the duodenum is an opening into which lead two conduits, or **ducts**, carrying digestive juices from two accessory organs of digestion. One of these juices is **pancreatic juice**, which arrives via the **pancreatic duct** from the **pancreas** (pan'kre-as). The other juice is **bile**, carried by the **common bile duct** from the **liver**. (Bile is not primarily a digestive juice in that it contains no enzymes, but it is important in the digestion of fats.) The small intestine secretes its own **intestinal juice**. Thus, as the food and the coin pass into the duodenum, they are deluged by a cloudburst of digestive juices.

Beyond the duodenum there are two more divisions of the small intestine: the **jejunum** (je-joo'num), which forms the next two fifths; and finally the **ileum** (il'e-um), constituting the remainder. The ileum joins the large intestine through another muscular valve called the **ileocecal** (il-e-o-se'kal) sphincter.

The food in the stomach is partially digested by the gastric juices, but the small intestine is the organ in which the greater part of the digestive process takes place, and where absorption occurs. We shall have a closer look at these two functions.

Absorption in the small intestine

Before the process of digestion is discussed in any detail, let us assume for the moment that the food in the small intestine has already been digested—that is, reduced by the concerted action of the digestive juices to the state in which it is ready to go to the blood stream and ultimately to the body cells. The means by which the digested food reaches the blood stream is known as **absorption**.

The small intestine is the chief organ of absorption, this process taking place through the mucosa by means of the countless minute projections of it known as villi. The villi are so small and so numerous that they give a velvety appearance to the lining of the small intestine. Each villus is of epithelium underlaid with connective tissue. Within each is a system of miniature arteries and veins, bridged with capillaries. All the basic food materials, including water and salts, but with the exception of some fats, are absorbed into the blood stream through the capillary walls in the villi. From here they pass by way of the portal system to the liver, to be stored or released and used as needed.

Fats have an alternate method of reaching the blood stream. As noted, some fat is absorbed via the blood capillaries of the villi. However, some fats also are absorbed by way of the lymphatic capillaries of the villi, which are called lacteals (these were mentioned in Chapter 11). According to some authorities about 40 percent of fat is absorbed via the lacteals. The word "lacteal" means "like milk," an apt description of the appearance of the mixture of lymph and fat globules that is drained from the small intestine after a quantity of fat has been digested. This mixture of fat and lymph, called chyle, collects in the cisterna chyli and eventually reaches the blood stream.

The process of digestion

In Chapter 3 on blood, it was pointed out that the blood plasma contains the water, the food substances and the mineral salts which together are necessary for the life and growth of the cells. We now know the means by which these materials reach the blood plasma in the first place; it yet remains for us to find out how these simple chemical substances are derived from the complex foods that we eat. This conversion process is, of course, known as digestion.

Let us review once more these basic materials which the cells need, and which are found in foods:

1. **Carbohydrates**, which include starches and sugars, and contain the elements carbon, hydrogen and oxygen.
2. **Fats**, which are more concentrated in fuel value than carbohydrates, and may, in some cases, carry important vitamins.
3. **Proteins**, which form the stuff of which, besides water, protoplasm is made.

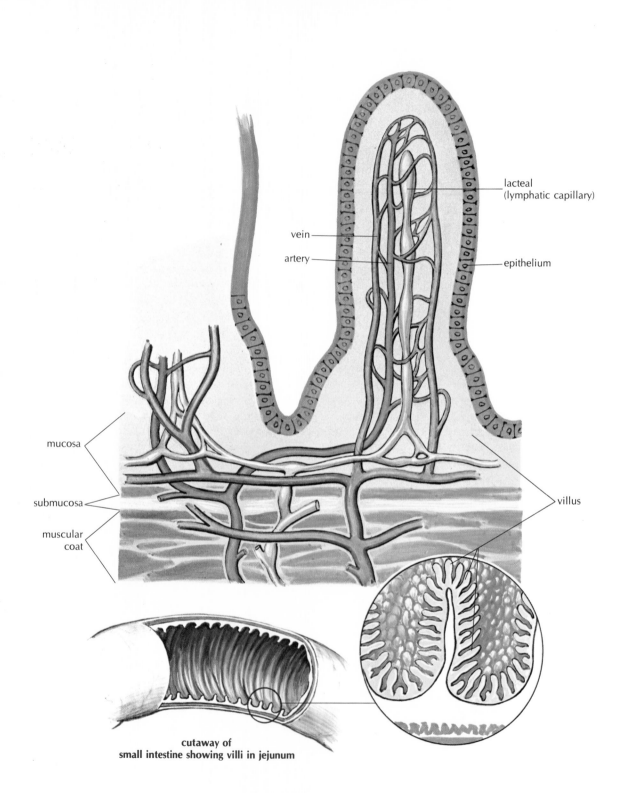

lacteal
(lymphatic capillary)

vein

artery

epithelium

mucosa

submucosa

muscular
coat

villus

cutaway of
small intestine showing villi in jejunum

Figure 12.4. Structure of a villus.

4. **Mineral salts**, which include a large variety of somewhat simpler compounds than those mentioned above. Salts maintain the proper conditions for osmosis in the cells, form a part of the body structure (as in bone) and play an important part in such life processes as muscle contraction and nerve responses.

5. **Vitamins**, which among other things help to regulate cell metabolism. They are food substances which are essential for good health.

6. **Water**, which constitutes about 66 percent of man's entire body composition. Eighty percent or more of many fruits and vegetables, and from 50 percent to 75 percent of meats and fish is water.

The digestive juices, by their chemical action, extract these various materials so that they can be absorbed. Most of the digestive juices contain the chemicals known as **enzymes**, and it is the enzymes that actually do the work of breaking down foods chemically. It should be pointed out that although the enzymes cause the chemical reaction to take place, they themselves do not enter into the reaction. For example, an enzyme may cause the basic protein to separate itself from the rest of the food, but the enzyme itself does not become a part of the protein. There are several different enzymes, and each acts on a specific food compound and no other. For example, some enzymes act only on fats; others act only on starches, and so forth. But let us see for ourselves what happens to a mass of food from the time it is taken into the mouth to the moment that it is ready to be absorbed.

The food is now being chewed; and saliva, the first of the digestive juices, is acting on it, softening it so that it can be swallowed easily. Saliva contains the enzyme **ptyalin** (ti'ah-lin), which initiates the process of digestion by changing some of the starches into sugars. Recall at this point that carbohydrates are found in the blood plasma in the form of simple sugar—glucose—and thus we should begin to understand where this blood sugar comes from.

When the food reaches the stomach, it is acted upon by the gastric juice, which contains hydrochloric acid and certain enzymes.

The most important functions of the gastric juice are those related to the actions of the hydrochloric acid (as previously outlined) and the liquefying of the food. In addition, there is some action, particularly on proteins, by the enzymes listed below:

1. **Pepsin**. This begins the breakdown of protein into simpler forms, so that the juices in the small intestine can work on them more effectively.

2. **Rennin**. This enzyme is the one that causes the curdling of milk, and is probably secreted in larger amounts in babies; possibly not at all in adults.

3. **Lipase** (lip'ace). This enzyme acts on fat but is of very minor importance in the stomach, partly because most of the fat particles found in the stomach are too large.

At this point the food is a liquid.(chyme) and proceeds to the small intestine for more chemical treatment.

Here the chyme is mixed with the greenish-yellow bile from the liver via the gallbladder. The bile does not contain enzymes; its action is purely mechanical. It works on fat, acting as a sort of liquid crowbar that splits the bits of fat into ever smaller particles so that the next digestive juice, the powerful secretion from the pancreas, can act more efficiently. The pancreatic juice contains a number of enzymes including the following:

1. **Lipase**. This enzyme was mentioned in connection with the gastric juice where it is present in small quantities. Following the physical division of fats into tiny particles by the action of bile, the powerful pancreatic lipase actually does almost all the digesting of fats. In this process fats are usually broken down into two simpler compounds, glycerol and fatty acids, which are more readily absorbable. If pancreatic lipase is absent, fats are expelled with the feces in undigested form.

2. **Amylopsin** (am-i-lop'sin). This changes starch to sugar.

3. **Trypsin** (trip'sin). This splits proteins into amino (am'e-no) acids, which are the form in which proteins enter the blood stream.

Digestive Juices and Enzymes

JUICES AND GLANDS	PLACE OF ACTION	ENZYMES	CHANGES IN FOODS
Saliva from 3 pairs of salivary glands	Oral cavity	Ptyalin	Begins starch digestion
Gastric juice from the stomach wall	Stomach	Pepsin Lipase Rennin	Begins protein digestion Digestion of fats Curdling of milk protein
Pancreatic juice from the pancreas	Small intestine	Amylopsin Trypsin Lipase	Acts on starches Acts on proteins Acts on fats
Intestinal juice from the small intestine (tubular glands)	Small intestine	Lactase Maltase Sucrase	Breaks down complex sugars into simpler forms
Bile from the liver	Small intestine	None	Breaks down fats physically so that lipase can digest them

The intestinal juice contains a number of enzymes including three that act on the sugars to transform them into the simpler form in which they are absorbed. These are **maltase, sucrase** and **lactase.** It must be emphasized that most of the chemical changes in foods occur in the intestinal tract because of the pancreatic juice, which could probably adequately digest all foods even if no other digestive juice were produced. If pancreatic juice is absent, serious digestive disturbances always occur.

This, in a nutshell, is the process of digestion. It should be noted that the food materials which the enzymes separate and reduce to absorbable forms are carbohydrates (i.e., sugars and starches), fats and proteins. The mineral salts are dissolved in the water, and this solution is absorbed as it is. The vitamins behave a bit differently according to their type. Some are incorporated in fats and are absorbed along with the fats (unless mineral oil is taken; and in that case the fat-soluble vitamins may be carried out in the feces). Other vitamins are dissolved in water and are absorbed in much the same way that mineral salts are. Still other vitamins (such as vitamin K) are produced by the action of bacteria in the colon and are absorbed from the large intestine.

THE LARGE INTESTINE

Once the processes of digestion and absorption have taken place, all that remains of the food is of no further use to the body, and so can be expelled. Also, anything else that may be indigestible (such as the coin that was swallowed) will pass out of the body through the large intestine.

The materials to be eliminated will continue through the exit or ileocecal valve from the small intestine and enter the small pouch at the beginning (proximal) part of the large intestine. This pouch is called the **cecum** (se'kum) and is located in the lower right iliac region of the abdomen. To the cecum is attached a small blind tube called the **vermiform** (ver'me-form) **appendix.** "Vermiform" means wormlike. The appendix contains relatively large amounts of lymphoid tissue such as that found in the tonsils; and as is frequent in the case of the tonsils, this tissue may become infected, a condition called **appendicitis.**

No enzymes are secreted by the large intestine. Its walls are lined with mucous membrane and contain layers of involuntary muscle which move the solid waste products, called **fecal matter,** on toward the rectum. Absorption of water takes place through the walls of the large intestine, though there are no villi. The action of bacteria within the large intestine is a further aid to the absorption of nutrients.

The colon and the rectum

The colon, which is the name for the longer part of the large intestine, extends up from the cecum along the right side and then bends to extend across the abdomen to the left side. It then descends on the left side of the abdomen into the pelvis. The lower part of the colon bends in an S shape to form the **sigmoid flexure** (sig'moid flek'sher). Hence the names for the four divisions: **ascending colon, transverse colon, descending colon** and **sigmoid colon** (see Fig. 12.1). The sigmoid colon empties into the **rectum**, which is about 6 to 8 inches long (15 to 20 cm.). The rectum serves as a temporary storage area for the indigestibles and unabsorbables. A narrow portion of the distal part of the large intestine is called the **anal canal**, which leads to the outside through an opening called the **anus** (a'nus). The whole journey from mouth to anus usually requires some 36 hours or more.

The Accessory Structures

THE LIVER
The liver, or **hepar** (he'par), is the largest of the glandular organs of the body. It is located under the dome of the diaphragm so that, if of normal size, it cannot be felt through the abdominal wall. The human liver is the same brownish-red color as the animal livers that you see in the market. It has a large right lobe and a somewhat smaller left lobe, as well as two other lesser lobes. It is a most remarkable organ, with so many functions that only a partial list can be made here. Some of these functions include:

1. The production of bile from the pigment of broken-down red blood cells.
2. The removal of poisons (toxins) that have been absorbed from the intestine.
3. The storage of simple sugar in a form called **glycogen**, which is released as needed in the form of glucose.
4. The final treatment of fats so that they can be more efficiently utilized by the cells.
5. The storage of certain vitamins, including A, D and some of the B group.
6. The manufacture of **heparin** (hep'ah-rin), an acid which prevents clotting of the blood.
7. The formation of antibodies, which act against disease organisms.
8. The production of certain blood plasma proteins such as fibrinogen and albumin.
9. The removal of a waste product called **urea** (u-re'ah) from amino acids.

THE GALLBLADDER
The gallbladder is a muscular sac that serves as a storage pouch for bile. While the liver may manufacture bile continuously, the need for it is likely to arise only a few times a day. Consequently, bile from the liver flows into the liver ducts and then up through the duct connected with the gallbladder. On the occasion when the semiliquid food mass (chyme) enters the duodenum, the gallbladder squeezes bile into the small bowel.

THE PANCREAS
The pancreas not only produces the pancreatic juice which we have noted; it also manufactures a substance called **insulin** which is released directly into the blood and has the function of regulating the amount of sugar which is "burned" in the tissues. We shall have a closer look at this function in Chapter 15.

Pancreatic juice, as mentioned, is extremely powerful; but of course remains harmless as long as it is confined to its proper channels.

The Peritoneum

The **peritoneum** is a serous membrane which covers the surface of most of the abdominal organs to form the visceral serosa and lines the abdominal wall to form the parietal layer. In addition to these parts of the peritoneum, there are more complex double layers of membrane which separate the abdomen into areas and spaces, and in some cases aid in supporting the organs and holding them in place (see Fig. 2.11).

The **mesentery** (mes'en-ter-e) is a double-layered peritoneal structure shaped somewhat like a fan, with the handle portion attached to the back wall. The expanded long edge is attached to the small intestine. Between the two layers of membrane that form the mesentery are the blood vessels, nerves and other structures which supply the intestine.

Another double-layered peritoneal structure, called the **greater omentum** (o-men'-tum), hangs downward from the lower border (greater curvature) of the stomach. This double layer of peritoneum extends into the pelvic part of the abdomen and then loops back and up to the transverse colon. It has been aptly described as an apron inside the abdomen. It may, in some cases, serve to prevent the spread of infection inside the abdominal cavity. There is also a peritoneal structure called the **lesser omentum**, which extends between the stomach and the liver.

Some Practical Aspects of Nutrition

Good nutrition is absolutely essential for the maintenance of health. This means basically that all the fundamental food materials necessary for the life and growth of the body cells must be continuously provided, in adequate quantity, in the food that we eat. If one or more of these vital materials are not supplied, the body will suffer in a number of ways, the effect being **malnutrition**, meaning bad nutrition. We commonly think of a malnourished person as one who does not have enough to eat; but malnutrition can occur just as easily from eating too much of the wrong foods. The normal manner by which malnutrition is avoided is to adhere to a balanced diet; that is, to insure that most of what is eaten includes adequate quantities of the basic nutrients.

In order that homemakers and others who plan meals may understand more easily how to provide a balanced diet, food groupings called the "Basic Seven" or the "Basic Four" have been publicized. Another division arranges foods in three groups as follows:

1. **Protective** foods, which are those especially high in vitamins and mineral salts. These include citrus and other fruits plus a variety of vegetables, particularly leafy green and yellow ones. Protective foods are especially valuable in staving off disease.
2. **Protein** foods, which are required for growth and repair of tissues. Since they cannot be stored, they should be included daily.
3. **Energy** foods, which contain fats and carbohydrates. They are needed in larger amounts by those who are especially active physically.

Summary

1. **Digestive system as a whole.**
   A. Functions: digestion and absorption.
   B. Components: alimentary canal, accessory organs.
2. **Alimentary canal.**
   A. Mouth, or oral cavity.
   (1) Structures.
      (a) Tongue.
      (b) Teeth: deciduous and permanent.
      (c) Salivary glands: parotid, submandibular (submaxillary), sublingual.
   (2) Diseases: gingivitis, stomatitis, pyorrhea, caries, mumps.
   B. Swallowing tubes and their accessories.
   (1) Pharynx (throat).
   (2) Uvula.
   (3) Epiglottis.
   (4) Esophagus ("gullet"). Has peristaltic action.
   (5) From swallowing tubes to anus, alimentary canal is lined with mucous membrane; has involuntary muscle for peristaltic action.

C. Stomach.
   (1) Characteristics and accessory structures.
      (a) Cardiac valve (guards entrance to stomach).
      (b) Rugae (folds in mucosa).
      (c) Stomach a storage pouch and churn (for chyme).
      (d) Juice contains hydrochloric acid and enzymes.
      (e) Pyloric valve (sphincter)—exit from stomach.
D. Small intestine.
   (1) Divisions: duodenum, jejunum, ileum.
   (2) Digestive juices and sources.
      (a) Intestinal juice (own secretion).
      (b) Bile from liver through common bile duct (no enzymes).
      (c) Pancreatic juice from pancreas via pancreatic duct.
   (3) Absorption: through villi.
      (a) Most food materials absorbed in blood capillaries.

(b) Some fats absorbed in lacteals (lymphatic).

    (4) Digestion: review table of enzymes.

E. Large intestine.

    (1) Has no enzymes. Water absorption occurs through walls.

    (2) Ileocecal valve (sphincter).

    (3) Cecum and appendix.

    (4) Colon (ascending, transverse, descending and sigmoid parts).

    (5) Rectum and anal canal, surrounded by anal sphincter.

3. Accessory structures.

A. Liver and its 9 major functions.

B. Gallbladder a storage pouch for bile; discharges bile when needed.

C. Pancreas.

    (1) Produces insulin (regulates use of sugar in cells).

    (2) Produces pancreatic juice (3 enzymes).

4. The peritoneum.

A. Serous membrane lining abdominal wall and covering abdominal organs.

B. Peritoneal structures: mesentery, greater omentum, lesser omentum.

5. Nutrition.

A. Balanced diet necessary for health.

B. Malnutrition: inadequate nutrition.

C. Three basic foods: protective, protein, energy.

Questions and Problems

1. Trace the path of an indigestible object from the mouth through all parts of the alimentary canal to the outside, and tell what happens on the way.

2. Differentiate between deciduous and permanent teeth as to kinds and numbers of the 2 sets.

3. What is the pharynx? What spaces connect with it?

4. How would you distinguish between "stomach" and "abdomen"?

5. What is peristalsis? Name some structures in which it occurs.

6. Name and describe the purposes of the acid in the stomach juice.

7. Describe the process of absorption.

8. What are the principal enzymes, and what is their origin? What does each do?

9. What is the mesentery and where is it located? The greater and lesser omenta?

10. In what ways are the tonsils and the appendix alike?

11. What constitutes a balanced diet?

The Respiratory System

Respiration

The word "respiration" means "to breathe again," and the fundamental purpose of the respiratory system is to supply oxygen to the individual tissue cells and to remove their gaseous waste product, carbon dioxide. Breathing—the popular term for respiration—refers merely to the inhaling and exhaling of air. Air is a mixture of oxygen, nitrogen, carbon dioxide, and other gases; the proportions of these gases varies depending upon the elevation above sea level and the amount of pollution in the specific locale. Respiration has two aspects; the first is that which takes place only in the lungs, where oxygen from the outside air enters the blood, and carbon dioxide is taken off from the blood to be breathed into the outside air. This aspect is **external respiration** (Fig. 13.1). The second aspect is called **internal respiration**. Internal respiration refers to the gas exchanges within the body cells. Oxygen leaves the blood and enters the cells at the same time that carbon dioxide leaves the cells and enters the blood. Internal respiration is also known as **cellular respiration** (see Fig. 10.8).

The respiratory system is an intricate arrangement of spaces and passageways which serve to conduct air into the lungs.

These spaces include the **nasal cavities**; the **pharynx**, which is common to the digestive and respiratory systems; the voice box, or **larynx** (lar'inks); the windpipe, or **trachea** (tra'ke-ah); and the **lungs** themselves, with their tubes and air sacs. The entire system might be thought of as a pathway for air between the atmosphere and the blood (Fig. 13.2).

The Respiratory System

NASAL CAVITIES

Air makes its initial entrance into the body through the openings in the nose called the **nostrils**. Immediately within are the two spaces known as the **nasal cavities**, located between the roof of the mouth and the cranium (the chamber that contains the brain). These two spaces are separated from each other by a partition, the **nasal septum**. The septum and the walls of the nasal cavities are constructed of bone covered with mucous membrane. At the side (lateral) walls of each nasal cavity are three projections called the conchae. The conchae greatly increase the surface over which the air must travel on its way through the nasal cavities.

The lining of the nasal cavities contains

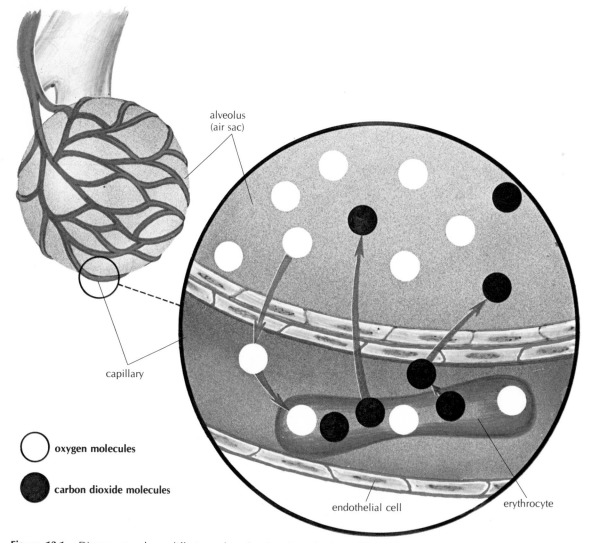

oxygen molecules

carbon dioxide molecules

alveolus
(air sac)

capillary

endothelial cell

erythrocyte

Figure 13.1. Diagram to show diffusion of molecules through the cell membrane and throughout the air in the alveolus and the capillary blood.

many blood vessels; hence it is described as a **vascular membrane**. The blood brings heat and moisture to the mucosa. As much as a quart of liquid secretion is believed to be produced daily by this membrane. The advantages of breathing through the nasal cavities over breathing through the mouth are due to the various changes effected on the air as it comes in contact with the parts of the nose, particularly the lining. These changes include:

1. The removal of foreign bodies, such as dust particles and pathogens, which

are either strained out by the hairs of the nostrils or caught in the surface mucus.

2. The warming of the air by the blood in the vascular mucosa.

3. The moistening of the air by the liquid secretion.

The sum of these changes amounts to a kind of air conditioning in a very real sense.

Also included in the discussion of the nasal cavities are the **sinuses**, which are small cavities, lined with mucous membrane, in the bones of the skull. The sinuses

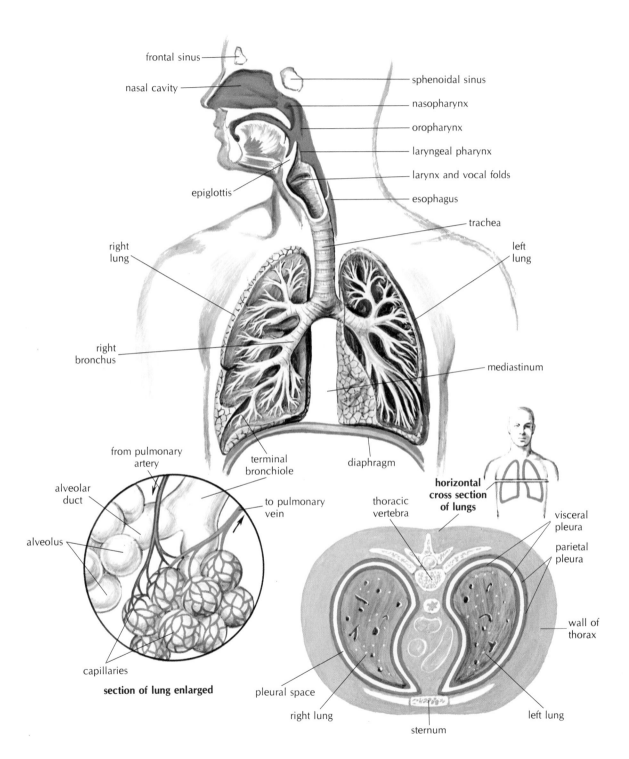

Figure 13.2. The respiratory system.

frontal sinus

nasal cavity

sphenoidal sinus

nasopharynx

oropharynx

laryngeal pharynx

larynx and vocal folds

esophagus

epiglottis

trachea

right lung

left lung

right bronchus

mediastinum

from pulmonary artery

terminal bronchiole

diaphragm

horizontal cross section of lungs

thoracic vertebra

alveolar duct

to pulmonary vein

visceral pleura

parietal pleura

alveolus

capillaries

section of lung enlarged

pleural space

right lung

sternum

wall of thorax

left lung

communicate with the nasal cavities, and they are highly susceptible to infection (see Chapter 6).

Another feature of the nasal cavities is a small duct communicating indirectly with the glands that produce tears. This is the **nasolacrimal** (na-zo-lak're-mal) **duct**, and its presence explains why the nose runs when tears flow freely.

The nasal cavities also contain the nerves and other structures which give us the sense of smell.

THE PHARYNX

The muscular pharynx serves as a passageway for foods and liquids into the digestive system, and for air into the respiratory tract. The upper portion located immediately behind the nasal cavity is called the **nasopharynx** (na-zo-far'inks). The middle section located behind the mouth is called the **oropharynx** (o-ro-far'inks); and finally the lowest portion is the **laryngeal** (lah-rin'je-al) **pharynx.** This last section opens into two spaces:

1. The air passageway into the larynx, which is toward the front.
2. The food path, toward the back (dorsally), entering the esophagus.

THE LARYNX

The **larynx**, or voice box, is located between the pharynx and the windpipe. It has a framework of cartilage which protrudes in the front of the neck and sometimes is referred to as the Adam's apple. The larynx is considerably larger in the male than in the female; hence the Adam's apple is much more prominent in the male. At the upper end of the larynx are the **vocal folds**. These cordlike structures serve in the production of speech. They are set into vibration by the flow of air from the lung. It is the difference in the size of the larynx that accounts for the characteristic male and female voices. Because a man's larynx is larger than a woman's, his voice is lower in pitch. The nasal cavities, the sinuses and the pharynx all serve as resonating chambers for speech, just as the cabinet does for a radio loudspeaker.

The space between these two vocal cords is called the **glottis** (glot'is), and the little leaf-shaped structure that closes this opening during swallowing is called the **epiglottis** (ep-e-glot'is). By the action of the epiglottis food is kept out of the remainder of the respiratory tract. The epiglottis acts as a lid or a trapdoor. As the larynx moves upward and forward during swallowing, the epiglottis moves downward, closing the opening into the larynx. During breathing, it rises to allow air to pass downward. Most of us are familiar with the choking sensation that occurs when we accidentally breathe and swallow at the same time.

The larynx is lined with ciliated mucous membrane. The cilia trap dust and other particles, moving them upward to the pharynx to be expelled, by sneezing or blowing the nose.

THE WINDPIPE, OR TRACHEA

The **trachea** is a tube that extends from the lower edge of the voice box to the center of the chest behind the heart. It has a framework of cartilage to keep it open. These cartilages, shaped somewhat like a tiny horseshoe or the letter C, are placed near each other along the entire length of the trachea. All the open sections of these cartilages are at the back so that the esophagus can bulge into this region during swallowing. The purpose of the trachea is to conduct air to and from the lungs.

THE BRONCHI

Near the center of the chest behind the heart the trachea divides into two **bronchi** (brong'ki). These are the two main air tubes entering the lungs, one on each side. The right bronchus is considerably larger in diameter than the left and extends downward in a more vertical direction. Therefore, if a foreign body is inhaled, it is likely to enter the right lung. Each bronchus enters the lung at a notch or depression called the **hilus** (hi'lus) or **hilum** (hi'lum). In this same region the blood vessels and the nerves also connect with the lung.

THE LUNGS

The **lungs** are the organs in which external respiration takes place; that is, where blood and air meet through the medium of the extremely thin and delicate lung tissues. There are two lungs, set side by side in the thoracic cavity, and each of them is constructed in the following manner:

As soon as each bronchus enters the lung at the hilum, it immediately subdivides. These branches or subdivisions of the bronchi resemble the branches of a tree, hence the common name, **bronchial tree.** Each individual bronchial tube subdivides again and again, forming progressively smaller divisions. The smallest are called **bronchioles** (brong'ke-oles). These tubes of assorted size contain small bits of cartilage which give firmness to the walls and serve to hold the tubes open so that air can pass in and out easily. However, as the tubes become smaller, the cartilage also decreases in amount until finally, in the most minute subdivisions, there is no cartilage at all.

At the end of each of the smallest subdivisions of the bronchial tree, called **terminal bronchioles,** there is a whole cluster of air sacs, resembling a bunch of grapes, known as **alveoli** (al-ve'o-li). Each air sac is made of one cell layer of squamous (flat) epithelium. This very thin wall provides an easy passage for the gases entering and leaving the blood which is contained in the millions of tiny capillaries of the alveoli. Some estimates indicate that there are some 700,000,000 of these alveoli in the human lung. The resulting surface in contact with gases approximates 73 square yards (60 square meters), about three times as much lung tissue as is necessary for life. Surely nature has allowed an ample margin of safety! Because of the many air spaces, the lung is light; and normally a piece of lung tissue dropped into a glass of water will float.

It will be recalled that the pulmonary circuit brings the blood to and from the lungs. The blood passes through the capillaries of the alveoli, where the gas exchange takes place.

The Lung Cavities

The lungs occupy a considerable portion of the chest (thoracic) cavity, which is separated from the abdominal cavity by the muscular partition known as the **diaphragm.** Each lung is enveloped in a sac of serous membrane called the **pleura;** hence there are two pleurae. The chest cavity is lined with this membrane also, this layer being known as the parietal pleura, while the lung covering is called the visceral pleura. Between the lungs is a space called the **mediastinum** (me-de-as-ti'num), containing the heart, among other things.

The entire thoracic cavity is flexible, capable of expanding and contracting along with the lungs. Its interior is well sealed off from the outside by its layer of membrane; and, as we shall see, this is a feature of the mechanism of breathing.

Physiology of Respiration

The mechanical process of breathing involves bóth the respiratory system and the muscles of respiration. The two phases of breathing, as mentioned earlier, are:

1. **Inhalation,** during which air is drawn into the lungs.
2. **Exhalation,** which refers to the expulsion of air from the alveoli.

Inhalation is the active phase of breathing, since it is then that the muscles of respiration, notably the diaphragm, contract in order to enlarge the chest cavity. The dome of the diaphragm is pulled downward, a partial vacuum is formed in the sealed pleural spaces, causing a pull on the elastic lung tissue so that air rushes in to fill the air sacs.

Exhalation is the inactive phase, since the muscles of respiration then relax, allowing the elastic tissues in the thorax to recoil and the abdominal organs to press upward against the diaphragm. Air is pushed outward by the weight of the rib cage and the upward boost of the abdominal viscera.

RESPIRATORY RATES

Normal rates of breathing vary from 12 to 25 times per minute, and are normally higher in children than in adults. The term **hyperpnea** (hi-perp-ne'ah) means "overbreathing due to abnormally rapid respiratory movements." **Apnea** (ap'ne-ah) means "a temporary cessation of breathing." It may be compensatory following forced respiration. In some fevers the respiratory rate increases in direct proportion to the increase in temperature, while in other cases there is no correlation between the respiratory rates and the temperature. It is important that the health worker check respiratory rates and record them properly. Observations should be done in such a way that the patient is unaware that a check is being made.

THE CONTROL OF BREATHING

Breathing is controlled by the respiratory center of the brain, which is located in the stem portion, called the **medulla**, immediately above the spinal cord. From this center, nerve fibers extend down into the spinal cord. From the neck part of the cord these nerve fibers continue through the **phrenic** (fren'ik) **nerve** to the diaphragm. Unlike the heart, the diaphragm does not continue to work if it is cut off from its nerve supply.

If one nerve is cut, the diaphragm on that one side is paralyzed.

The diaphragm and the other muscles of respiration are voluntary in the sense that they can be regulated by messages from the higher brain centers. It is possible for a person deliberately to breathe more rapidly or more slowly, or to hold his breath and not breathe at all for a time. Usually we breathe without thinking about it, while the respiratory center in the medulla does the controlling. This center is governed by variations in the chemistry of the blood. If there is an increase in carbon dioxide in the blood, the cells of the respiratory center are stimulated; and they in turn send impulses down the phrenic nerves to the diaphragm.

Summary

1. **Respiration.**
   A. Purpose: supply oxygen to tissues, remove carbon dioxide.
   B. Aspects.
      (1) External: gas exchange in lungs.
      (2) Internal: gas exchange in tissues.
2. **Respiratory system as a whole.**
   A. Nasal cavities: include sinuses, nasolacrimal duct.
   B. Pharynx (throat): passage way for both food and air.
   C. Larynx (voice box).
   D. Trachea (windpipe): conducts air to bronchi, lungs.
   E. Bronchi: 2 tubes branching at end of trachea; each to a lung.
   F. Bronchial tubes: subdivisions of bronchi in lungs.
   G. Bronchioles: smallest subdivisions of bronchial tree.
   H. Alveoli: air sacs where gas exchange occurs. Connected to terminal bronchioles. Contain capillaries of pulmonary circulation.
3. **Lung cavities.** Lungs occupy most of thoracic cavity. Both lungs covered by, and thoracic cavity lined with, pleura. Diaphragm separates thoracic and abdominal cavities. Space between lungs is mediastinum, containing heart and other organs.
4. **Physiology of respiration.**
   A. Phases of breathing: inhalation, exhalation.
   B. Rate: 12 to 25 times a minute.
   C. Control: by respiratory center in medulla, through phrenic nerves to diaphragm.

Questions and Problems

1. What is the purpose of respiration, and what are its 2 aspects?
2. Trace the pathway of air from the outside into the blood.
3. What are the advantages of breathing through the nose?
4. Locate and describe: nasal septum, nasolacrimal ducts, mediastinum, phrenic nerves.
5. Describe the lung cavities.
6. Describe normal breathing, including: 2 phases, respiratory rates, mechanism of breathing and nerve control.

The Urinary System

The **urinary system** is also called the **excretory system**, and its job is to remove certain waste products from the blood and eliminate them from the body (Fig. 14.1).

What does the name of this system mean? The terms "elimination" and "excretion" often are used interchangeably. Actually, we usually think of **excretion** as the function of removing useless substances (i.e., the waste products of cell metabolism) from the blood and the lymph, often by a type of cellular activity similar to that of gland cells producing a **secretion**. For example, we may speak of the secretory function of the kidney cells along with their excretory function. Urine, then, is both a secretion and an excretion. On the other hand, **elimination** indicates the actual emptying of the hollow organs in which these waste substances have been temporarily stored. The kidney would then excrete, while the urinary bladder eliminates. In the digestive system the liver would excrete bile (again by a process of secretion) while the large intestine would eliminate this bile in the feces.

Here it may be of interest to summarize the chief excretory mechanisms of the body, along with the various substances which they eliminate:

1. The urinary system: water, waste products containing nitrogen and salts. These are all contained in the urine.
2. The digestive system: water, some salts, bile and the residue of digestion. These are all contained in the feces.
3. The respiratory system: carbon dioxide and water. The latter appears as vapor, as breathing on a windowpane will demonstrate.
4. The skin, or integumentary system: water, small quantities of nitrogenous wastes and salts. These all appear in perspiration, though evaporation of water from the skin may go on most of the time without our being conscious of it.

This chapter will, of course, cover only the first of these systems. However, when we come to such a subject as the means by which the body maintains its delicate balance of water and various chemicals, other systems pertinent to the discussion will be

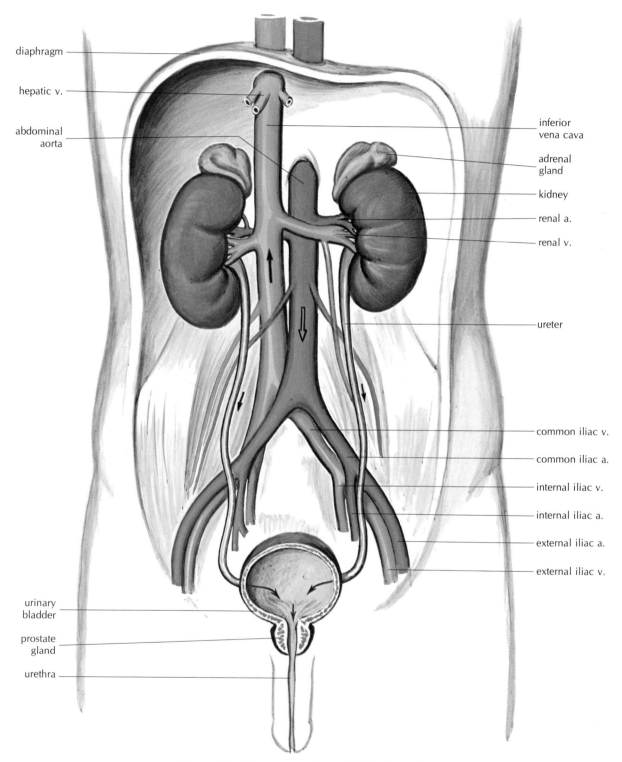

Figure 14.1. The urinary system, with blood vessels.

referred to. No system of the body operates independently of the others, after all, in spite of its specialized functions. A study of the urinary system will reveal several of these interrelationships.

Organs of the Urinary System

The main parts of the urinary system are:
1. Two **kidneys**, which are glandular organs that are necessary for life. These are the organs which, in addition to other things, extract wastes from the blood.
2. Two **ureters** (u-re'ters), which conduct the secretion from the kidneys to the urinary bladder.
3. The single **urinary bladder**, a reservoir which receives the urine brought into it by the two ureters.
4. A single **urethra** (u-re'thrah), which is the excretory tube for the bladder. Through the urethra the urine is conducted to the outside of the body.

The Kidneys

LOCATION OF THE KIDNEYS
The two kidneys lie against the muscles of the back in the upper abdomen. They are protected by the ribs and their cartilages, since they are up under the dome of the diaphragm. Each kidney is enclosed in a membranous capsule that is made of fibrous connective tissue; it is loosely adherent to the kidney itself. In addition there is a circle of fat around the perimeter of the organ. It is called the adipose capsule and is one of the chief supporting structures. The peritoneum covers the front of the kidneys only, and so these and several other structures are not in the peritoneal cavity. This area is known as the **retroperitoneal** (re-tro-per-it-o-ne'al) **space**, indicating that it is behind the peritoneum (see Fig. 2.11.).

STRUCTURE OF THE KIDNEYS
The kidney is a somewhat flattened organ about 4 inches (10 centimeters) long, 2 inches (5 centimeters) wide, and 1 inch (2.5 centimeters) thick. On the inner or medial border there is a notch called the **hilum** (hi'lum), at which region the artery, the vein and the ureter connect with the kidney. The outer or lateral border is convex (curves outward), giving the entire organ a bean-shaped appearance (Fig. 14.2).

The kidney is a glandular organ; that is, most of the tissue is epithelium with just enough connective tissue to serve as a framework. As is the case with most organs, the most fascinating aspect of the kidney is too small to be seen with the naked eye. This basic unit of the kidney, where the kidney's business is actually done, is called a **nephron** (nef'ron); and a nephron is primarily a tiny coiled tube (called a **convoluted tubule**) with a bulb on one end containing a cluster of capillaries, the **glomerulus** (glo-mer'u-lus) (Fig. 14.3A and B). A kidney is composed of over a million nephrons; if all these coiled tubes were separated, straightened out and laid end to end, they would span some 75 miles (120 kilometers)!

The cluster of capillaries within the bulb of each nephron has one blood vessel (an arteriole) to supply it with blood, and another tiny vessel to drain it. As the blood passes through the capillary cluster, a mixture of water, useful materials and dissolved waste materials passes directly from the blood (by filtration), through the capillary walls, and into **Bowman's capsule**, the beginning of the convoluted tubule (see Fig. 14.4). Now this mixture begins its journey through the coiled tube of the nephron. As it passes through the tube, most of the water, minus the waste materials but containing the useful substances that have escaped through the nephron capillaries, is *reabsorbed* through the walls of the nephron tube and sent back to the blood stream. Thus we see that each nephron is able to "clean" or filter a very large volume of blood without causing the body to lose too much of its water or other essential materials. The water that the nephrons do retain becomes increasingly more concentrated with waste materials; and this concentrated mixture now is known as **urine**.

The nephron bulbs, with their blood vessels, are located in the **cortex**, or outer part of the kidney. The open distal ends of the nephron tubes come together at the **collecting tubules**, located within the **medulla**, or inner part of the kidney. Inside the kidney, toward the medial part, the ureter expands to form a basin which receives the urine col-

159

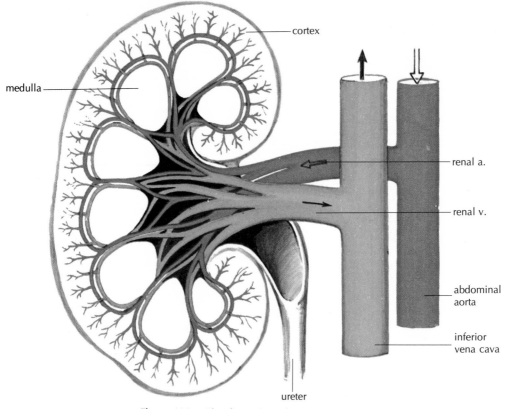

cortex

medulla

renal a.

renal v.

abdominal
aorta

inferior
vena cava

ureter

Figure 14.2. Blood supply and circulation of kidney.

lected by the collecting tubules in the medulla. This space is referred to as the **renal basin** (or, sometimes, the **pelvis** of the kidney). In order to increase the area for collection, tubelike extensions project from the renal pelvis into the active kidney tissue. These extensions are called **calyces** (kal′i-sez). The urine which collects in the renal basin passes down the ureters to the bladder (Fig. 14.5).

THE KIDNEYS AND BODY CHEMISTRY

The kidneys have three main functions. The first, which has been discussed already to some extent, is **excretion**. The use of proteins by the body cells (in the form of amino acids) produces, among other waste materials, those containing the chemical element nitrogen; the chief waste product of this category is **urea** (u-re′ah). The urinary system is the specialized mechanism of excretion for this nitrogenous waste material. Certain salts from the blood plasma are also excreted.

A second function of the kidneys is to aid in the maintenance of **water balance**. The average man takes in about 2½ quarts (about 2500 cc.) of water daily. About half of this usually comes from foods, some of which contain considerable amounts of water, such as fruits, many vegetables, soups and milk. In addition to this, more than a tumblerful (about 300 cc.) of water is formed in the cells when their materials combine with oxygen. On the other hand, water is constantly being lost in a number of ways. About a quart and a half (1500 cc.) is lost through the urine each day. Considerable water is lost in fecal material, and every exhalation is accompanied by water loss. Even though the weather may be cool, some moisture is lost in the form of unnoticed perspiration. In spite of this great variation in the amount of water which the body takes in and gives off, the water in the tissues must be maintained at a constant level. The mechanism for accomplishing this is complicated, but the kidneys are an important part of it. It

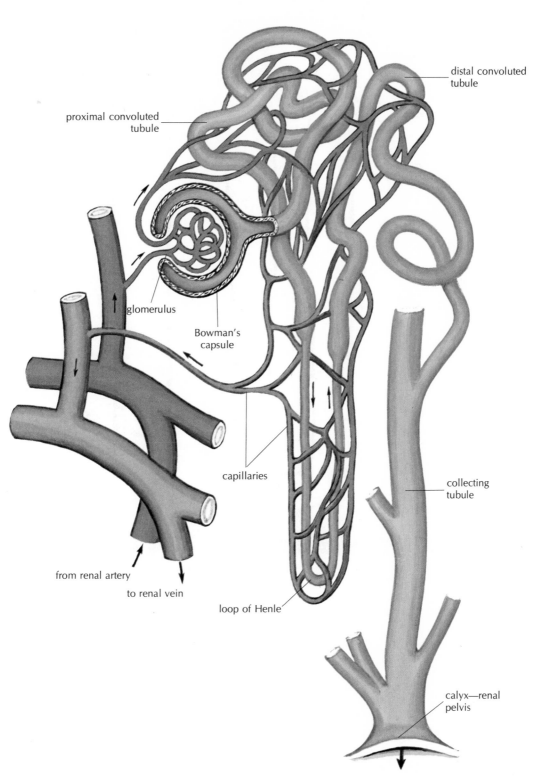

proximal convoluted tubule

distal convoluted tubule

glomerulus

Bowman's capsule

capillaries

collecting tubule

from renal artery

to renal vein

loop of Henle

calyx—renal pelvis

Figure 14.3A. A simplified nephron.

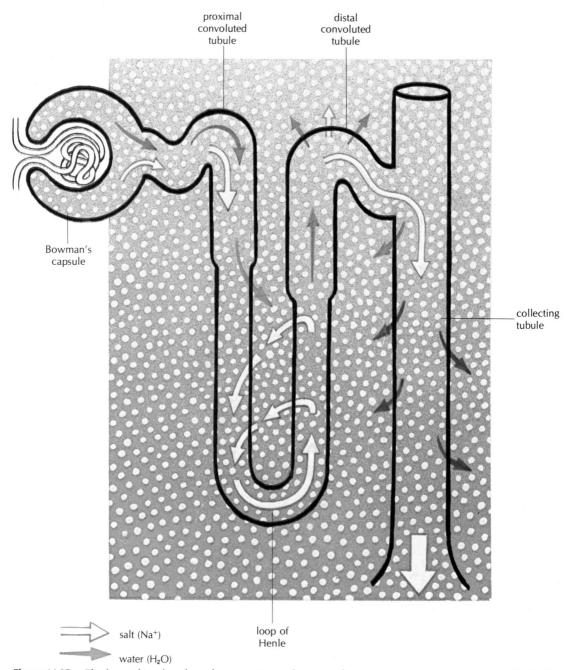

proximal convoluted tubule

distal convoluted tubule

Bowman's capsule

collecting tubule

loop of Henle

⇨ salt (Na⁺)

➡ water (H₂O)

Figure 14.3B. The loop of Henle, where the proportions of waste and water in urine are regulated according to the body's constantly changing needs. The concentration of urine is determined by means of intricate exchanges of water and salt.

may be noted that the kidneys serve as a sort of "overflow" for water that the body does not need. If, for example, a person deliberately drinks a large amount of water even though he may not feel thirsty, most of this water will very quickly be excreted by the kidneys; and furthermore this water will contain only a small concentration of waste materials.

A third function of the kidneys is to aid in

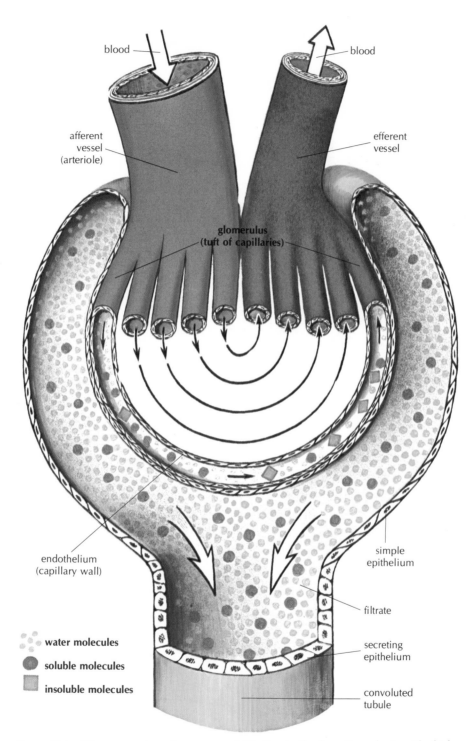

blood

blood

afferent
vessel
(arteriole)

efferent
vessel

**glomerulus
(tuft of capillaries)**

endothelium
(capillary wall)

simple
epithelium

filtrate

water molecules

soluble molecules

insoluble molecules

secreting
epithelium

convoluted
tubule

Figure 14.4. Diagram to show the process of filtration in the formation of urine. The higher pressure inside the small capillaries of the glomerulus forces the dissolved substances (except the plasma proteins) and much water into the space inside Bowman's capsule. The smaller caliber of the efferent vessel as compared with the larger afferent vessel causes this higher pressure.

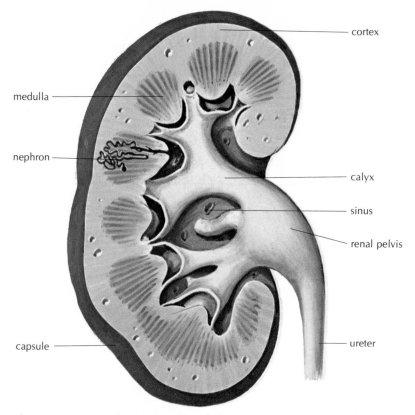

Figure 14.5. Longitudinal section through the kidney showing its internal structure.

regulating the **acid-base balance** of the body. Acids are a category of chemical substances which, in the body, are produced by cell metabolism. They may take the form of solids, liquids or gases. Bases, also called alkalies, are another category of chemicals; these have the effect of neutralizing acids (the product of an acid-base reaction is a salt). Certain foods can cause acids or alkalies to form in the body. In order that all the normal body processes may take place, a certain critical proportion of acids and bases must be maintained at all times—this despite the fact that the person may take in varying quantities of acid-forming and alkaline-producing substances as food. It is just as important for the blood and other tissues that an excess of alkalinity be avoided as it is that too much acidity be prevented. In other words, there must be a balance. Acid substances are constantly being removed from the body in various ways, including the exhalation of carbon dioxide, which serves to remove carbonic acid. Both acid and alkaline substances which may be present in excess are constantly being removed by the kidneys. The kidneys are also able to manufacture ammonia at certain times. Ammonia neutralizes acids, and so we have another example of the kidneys' ability to help maintain the acid-base balance.

An interesting and equally important sidelight on the maintenance of the acid-base balance is the presence in the blood of certain mineral salts called "buffers," which maintain the blood in its nearly neutral, slightly alkaline state in spite of the varying character of the food that is taken in.

The Ureters

The two ureters are long, slender, muscular tubes that extend from the kidney basin down to and through the lower part of the urinary bladder. Their length naturally varies with the size of the individual, and so may be anywhere from 10 to 13 inches long (25 to

32 cm.), including nearly an inch (2.5 cm.) at its lower part that passes obliquely through the bladder wall. They are entirely extraperitoneal; that is behind and, at the lower part, below the peritoneum.

The wall of the ureter includes a lining of epithelial cells, a relatively thick layer of involuntary muscle, and finally an outer coat of fibrous connective tissue. The muscle of the ureters is capable of the same rhythmic contraction found in the digestive system and known as peristalsis. Urine is moved along the ureter from the kidneys to the bladder by peristalsis at frequent intervals. Because of the oblique direction of the last part of each ureter through the lower bladder wall, compression of the ureters by the full bladder prevents backflow of urine.

The Urinary Bladder

The urinary bladder is located below the parietal peritoneum and behind the pubic joint when it is empty. When it is filled, it may extend well up into the abdominal cavity proper. The urinary bladder is a temporary reservoir for urine, just as the gallbladder is a storage bag for bile.

The bladder wall has many layers. It is lined with mucous membrane; the lining, like that of the stomach, is thrown into the folds called rugae when the receptacle is empty. Beneath the mucosa is a layer of connective tissue. Then follows a three-layered coat of involuntary muscle tissue which is capable of stretching to a great extent. Finally there is an incomplete coat of peritoneum that covers only the upper portion of the bladder. When the bladder is empty, the muscular wall becomes thick and the entire organ feels firm. As the organ fills, the muscular wall becomes thinner and the organ may increase from a length of 2 or 3 inches to 5 or more inches (5 to 12.5 cm.). A moderately full bladder holds about a pint of urine (470 cc.).

Near the outlet of the bladder circular muscle fibers contract to prevent emptying, and form what is known as the **internal sphincter**. In the infant a center in the lower part of the spinal cord receives impulses from the bladder and sends motor impulses out to the bladder musculature; the organ is emptied in an action that is automatic (i.e., a **reflex** action). However, with training, the child learns to control this reflex.

The Urethra

The **urethra** is the tube that extends from the bladder to the outside, and is the means by which the bladder is emptied. The urethra differs in men and women, since in men it is also a part of the reproductive system and it is much longer.

The female urethra is a thin-walled tube about 1½ inches (3.75 cm.) long. It is behind the pubic joint and is embedded in the muscle of the front wall of the vagina. The external opening is called the **urethral meatus** and is located just in front of the vaginal opening.

The male urethra is about 8 inches long. In the first part of its course it passes through the prostate gland, where two ducts carrying the male sex cells join it. From here it leads through the **penis** (pe'nis), the male organ of copulation, to the outside. The male urethra, then, serves the dual purpose of conveying the sex cells and draining the bladder, while the female urethra performs only the latter function.

The process of expelling urine through the urethra is called **urination** or **micturition**. It is controlled by the action of circular muscles continuous with those in the walls of the bladder and in the urethra. These form valvelike structures that are aided by external muscles in the pelvic floor.

The Urine

In this chapter we have learned some of the main constituents of urine. Here they are summarized in a more detailed manner in order to complete the picture of the normal situation.

Urine is a yellowish liquid which is about 95 percent water. Dissolved in this water are a number of solids and a few gases, including:

1. Nitrogen waste products, which include urea, uric acid and **creatinine** (kre-at'i-nin). These substances sometimes are classified as **organic** com-

pounds because they are formed by the breakdown of cells in the body and from proteins of food, many of which originated from animal tissues. That is, they originated from living organisms.

2. Mineral salts, including compounds such as sodium chloride (as in common table salt), sulfates and phosphates of different kinds. These substances are often classified as in-

organic compounds because they do not originate in living organisms. They are excreted in appropriate amounts in order to keep the blood concentrations of mineral salts constant.

3. Yellow pigment, which is derived from certain bile compounds. The color of urine varies with the concentration, being lighter when more dilute, as a rule.

Summary

1. **Excretion and elimination.**
   A. Excretion: removal of useless substances from the body structures.
   B. Elimination: emptying of organs in which waste products have been stored.
   C. Excretory mechanisms: digestive, urinary, respiratory, skin.
2. **Urinary system:** kidneys, ureters, bladder, urethra.
3. **Kidneys.**
   A. Location: retroperitoneal.
   B. Structure: largely epithelium in tiny tubes (nephrons) acting as filters. Nephrons lead to collecting tubules in medulla. Urine from collecting tubules collected in renal basin (pelvis) where ureters join.
   C. Functions: excretion; maintenance of water balance; regulation of acid-base balance.

4. **Ureters.**
   A. Structure and function: muscular tubes that carry urine from kidney to bladder by peristalsis.
5. **Bladder.**
   A. Structure and function: muscular sac capable of stretching; reservoir for urine. Has internal sphincter; impulse to empty a controllable reflex.
6. **Urethra.**
   A. Structure and function: tube leading from bladder to outside; longer in male than in female; in male is also part of reproductive system.
7. **Urine.**
   A. Normal constituents: water (95 percent), nitrogenous wastes; mineral salts; pigment.

Questions and Problems

1. In what way might secretions and excretions be the same, and how do these differ from each other?
2. How do the terms "elimination" and "excretion" differ from each other? Name the body systems which have an excretory function.
3. Where are the kidneys located?
4. Describe the external appearance of the kidneys and tell what tissues form most of the kidney structure. Name and describe the microscopic structure that is the basic kidney unit.
5. What structures empty into the kidney pelvis, and what drains this space?
6. In what 3 ways does the kidney adjust the body chemistry?
7. Locate and describe the ureters.
8. Name and describe the layers of the bladder wall.
9. Describe the female urethra and tell how it differs from the male urethra in structure and function.
10. What are some of the most important normal constituents of urine?

Glands and Hormones

Classification of Secretions and Glands

A gland as such is any organ that produces a certain secretion; the secretions themselves are substances manufactured from blood constituents by the specialized cells of which the glands are made.

The secretions of the various glands may be divided into two main groups:

1. **External secretions**, which are carried from the gland cells to a nearby organ or to the body surface. These external secretions are effective in a limited area near their origin. We have already learned something of most of these secretions: digestive juices, the secretions from the sebaceous glands of the skin, tears from the lacrimal glands and urine (both a secretion and an excretion).

2. **Internal secretions**, which are carried to all parts of the body by the blood or lymph. These substances often affect tissues a considerable distance from the point of origin. Internal secretions are known as **hormones**; and the hormones, with the glands that produce them, will be the subject of this chapter.

Glands also fall into two categories:

1. **Exocrine** (ek'so-krin) **glands**, which have tubes (ducts) to carry the secretion from the gland to another organ or part of the body.

2. **Endocrine** (en'do-krin), or **ductless**, **glands**, which have no ducts, and so must depend upon the blood and the lymph to carry their secretions to various body tissues, via the capillaries of the glandular tissue (Fig. 15.1).

Sometimes the lymph nodes are spoken of as glands (i.e., the "neck glands," which

167

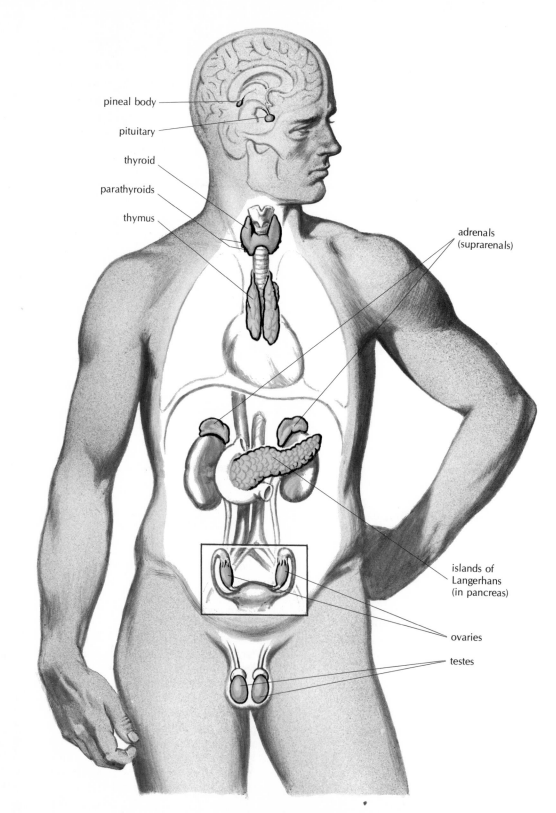

pineal body

pituitary

thyroid

parathyroids

thymus

adrenals
(suprarenals)

islands of
Langerhans
(in pancreas)

ovaries

testes

Figure 15.1. Endocrine system, with female sex glands shown in inset.

are actually nodes); but so far as is known the lymph nodes do not produce secretions.

The endocrine glands proper produce only internal secretions (hormones), but some organs contain both exocrine and endocrine gland tissue. The liver and the pancreas, for example, are combination glands, producing both external and internal secretions. The mucous membranes of the stomach and the small intestine produce both external secretions (mucus, digestive juices) and a hormone called **secretin** (se-kre′tin), which has the function of stimulating certain other digestive organs to produce *their* digestive juices.

Structure of Glands

Exocrine glands vary in complexity from very simple depressions resembling tiny dimples to involved arrangements such as are found in the kidneys. Simple tubelike structures are found in the stomach wall and in the intestinal lining. Complex, treelike groups of ducts are found in the liver, the pancreas and the salivary glands. Most glands are made largely of epithelial tissue with a framework of connective tissue. There may be a tough connective tissue capsule (a fibrous envelope) enclosing the organ, with extensions into the organ which form partitions. Between the partitions are groups of cells, and these units are called **lobes**.

Most endocrine glands, like exocrine glands, are made of epithelial tissue. However, since they have no ducts, they seem to make up for this lack by a most extensive blood vessel network. Operations on endocrine glands such as, for example, on the thyroid, require care in the control of bleeding. The organs believed to have the very richest blood supply of any in the body are the tiny **adrenal**, or **suprarenal**, **glands** which are located near the upper parts of the kidneys.

General Functions of Hormones

Hormones are chemical substances manufactured by the endocrine glands, and their overall function is to regulate the activities of various body organs. In one sense, the action of hormones can be compared with that of the nervous system (in fact, in the chapter covering the nervous system, we saw how the hormone epinephrine acted in conjunction with the autonomic system in an emergency). Hormones sometimes are referred to as "chemical messengers." Some hormones stimulate exocrine tissues to produce their secretions, as in the example of secretin. Another category of hormone stimulates other endocrine glands to action. There are hormones that have a profound effect upon growth, development, and even the personality of the individual. Others regulate the body chemistry: for example, the metabolism of the cells. Still others regulate the contraction of muscle tissues. Finally, there are hormones which control various sex processes. The action of each hormone is specific; that is, each has its particular, specialized job to do and no other.

Endocrine Glands and Their Hormones

THE THYROID GLAND AND THYROXINE

The largest of the endocrine glands is the **thyroid**, which is located in the neck. It has two oval parts called the lateral lobes, one on either side of the voice box. A narrow band called the **isthmus** (is′mus) connects these two lobes. The entire gland is enclosed by a connective tissue capsule. This gland produces the hormone **thyroxine** (thi-rok′ sin). The principal function of thyroxine is to regulate metabolism for the production of heat and energy in the body tissues. In order that this hormone may be manufactured, there must be an adequate supply of iodine in the blood. The iodine content may be maintained by eating vegetables grown in iodine-containing soils, or by eating sea foods. In some parts of the United States, as well as in certain regions of Italy and Switzerland, the soil is so deficient in iodine that serious defects may result in the inhabitants of these districts. The use of iodized salt may prevent difficulties.

Thyroxine causes an increase in enzyme activity within each and every body cell.

Some of the effects of this include:

1. Increased absorption and utilization of glucose.
2. Increased use of fat because glucose is depleted earlier in the presence of thyroxine.
3. Decreased quantity of fats (including cholesterol) in the circulating blood.
4. Accelerated growth and repair of body tissues because of the stimulation of protein metabolism, the anabolism (building up) phase, as well as increased catabolism (breaking down) by which proteins are used as fuel.
5. Greater need for vitamins because vitamins are required for enzyme activity.
6. Accelerated heart rate and increased strength of the heart beat.
7. Increased rate and depth of respiration.
8. Increased muscular and nervous activity.

Tests for thyroid function

The two tests most commonly used in evaluating thyroid function are:

1. The **basal metabolism test**, in which the amount of oxygen a person uses is determined while the individual is at rest. It must be carefully supervised and be performed in the morning before breakfast. It is preferable that a person have this test done after having stayed in the hospital overnight, since conditions can be more closely controlled by hospital personnel prior to the test.
2. The **blood iodine test**, in which blood is taken from a vein and then tested for the amount of so-called protein-bound iodine present. This is a much simpler test, and in most cases is very helpful in determining the activity of the thyroid.

A third test involves the use of radioactive iodine, in which excretion of iodine is studied following the ingestion of a measured amount. The collection of every specimen of urine is mandatory if the test is to be accurate. Radioactive substances are used also in the treatment of some cases of hyperthyroidism.

THE PARATHYROID GLANDS AND PARATHORMONE

Behind the thyroid gland, and embedded in its capsule, are four tiny epithelial bodies called the **parathyroid glands.**

The hormone produced by these glands regulates the amount of calcium dissolved in the circulating blood. If these glands are removed, there will follow a series of muscle contractions involving particularly the hand and the face muscles. These spasms are due to a low concentration of blood calcium, and the condition is called **tetany** (tet'ah-ne), which must *not* be confused with the infection called tetanus (lockjaw). On the other hand, if there is an excess production of the secretion of these glands, called **parathormone** (par-ah-thor'mone), as may happen in tumors of the parathyroids, calcium (normally stored in the bones for use by the tissues as needed) is removed from its storage place and is poured into the blood stream, whence it finally is excreted by the kidneys. Because of the outpouring of calcium from the bones, they become soft and easily bent. It is believed that the excess of calcium in the blood is one cause of kidney stones.

THE PITUITARY, OR "MASTER GLAND"

The **pituitary** (pi-tu'i-tar-e) is a small gland about the size of a cherry. It is nearly surrounded by bone except for its area of connection with the brain. The pituitary is located in a saddlelike depression just behind the point of optic nerve crossing, in the midline. It has two important parts called the **anterior** and the **posterior lobes**, each of which produces several different hormones.

The anterior lobe

This lobe produces a large number of hormones. Many of them stimulate other glands, and it is for this reason that the pituitary is known as the master gland. Its main hormones are:

1. The **growth-promoting**, or **somatotropic** (so-mah-to-trop'ik), **hormone**. If a human is born with a deficiency of this hormone, he will remain a dwarf (of the type often referred to as a midget).
2. The **gonadotropic** (gon-ad-o-trop'ik) **hormones**, which control the development of the reproductive systems in

The Endocrine Glands and Their Hormones

GLANDS	HORMONES	PRINCIPAL FUNCTIONS
anterior pituitary (adenohypophysis)	ACTH adrenocorticotrophin	stimulates adrenal cortex, which aids in protecting the body in stress situations (injury, pain, etc.)
	TH or TSH (thyroid stimulating)	stimulates thyroid gland to produce thyroxine
	FH or FSH (follicle stimulating)	stimulates growth and hormone activity of ovarian follicles
	HGH (human growth hormone)	promotes growth of all body tissues
	LH (luteinizing; a gonadotropin)	causes development of corpus luteum at site of ruptured ovarian follicle
	LTH (luteotropic; a gonadotropin)	stimulates hormone activity of luteal cells, which produce progesterone
posterior pituitary (neurohypophysis)	ADH (antidiuretic; or vasopressin)	promotes reabsorption of water in kidney tubules; stimulates involuntary muscle tissue
	oxytocin	causes contraction of pregnant uterus
adrenal cortex	cortisol (95% of glucocorticoids)	aids in metabolism of food components (proteins, carbohydrates and fats); active during stress
	aldosterone (95% of mineralocorticoids)	aids in regulating mineral salts (electrolytes)
adrenal medulla	epinephrine and norepinephrine	increases blood pressure and heart rate, activates cells influenced by the sympathetic nervous system plus many not affected by sympathetics
pancreatic islets	insulin	required for glucose metabolism, for the use of glucose within body cells
	glucagon	maintains amount of glucose in blood adequate for brain and other organs to which glucose is so vital
parathyroids	parathormone	increases the amount of calcium ion in the blood
prostaglandins (?) various	cyclic fatty acids	believed to act on central nervous, cardiovascular and gastrointestinal systems
thyroid gland	thyroxine	increases metabolic rate, influencing both physical and mental activities; required for normal growth and catabolism
	calcitonin	decreases blood level of calcium
	triiodothyronine	potent activator of all metabolic activities
ovarian follicle	estrogen	stimulates growth of primary sexual organs (uterus, tubes, etc.) and development of secondary sexual organs, such as breasts, plus change in pelvis to ovoid broader shape
corpus luteum (in ovaries)	progesterone	stimulates development of secretory parts of mammary glands; prepares uterine lining for implantation of fertilized ovum; aids in maintaining pregnancy; enhances reabsorption of water and salt (NaCl) from kidney tubules
testes	testosterone	stimulates growth and development of sexual organs (testes, penis, etc.) plus development of secondary sexual characteristics such as hair growth on body and face, deepening of voice, and increase in development and strength of muscles
placenta	chorionic gonadotropin	causes continued growth and secretory activity of corpus luteum (of ovary)
	estrogens	stimulate growth of mother's reproductive organs, cause relaxation of pelvic ligaments (to make childbirth easier)
	progesterone	aids in maintaining nutrition of embryo (due to effects on uterine lining cells)
stomach lining (mucosa)	gastrin	stimulates secretion of HCl and some enzymes
intestinal mucosa	prosecretin (becomes secretin)	activates the pancreas to produce an alkaline watery pancreatic juice (to neutralize HCl), and stimulates intestinal glands and liver to produce their digestive juices
	cholecystokinin	cooperates with peristalsis of small intestine to cause contraction and emptying of gallbladder

both the male and the female. This action includes the stimulation of the monthly menstrual cycle of the female.

3. The **thyrotropic** (thi-ro-trop'ik) **hormone**, which stimulates the thyroid gland to activity.
4. The **adrenocorticotropic** (ah-dre-no-kor-ti-ko-trop'ik) **hormone**, abbreviated ACTH, which stimulates the cortex of the adrenal gland (to be discussed in this chapter).
5. The **lactogenic** (lak-to-jen'ik) **hormone**, which stimulates the production of milk in the female.

The posterior lobe

This lobe of the pituitary gland stores and releases three hormones. They are apparently produced by the hypothalamus, from which they flow down nerve fibers to the posterior pituitary. Their functions are:

1. To stimulate the smooth muscle of the blood vessels.
2. To promote reabsorption of water in the tubules of the kidney.
3. To stimulate the contraction of the muscles of the uterus.

An extract containing these hormones, called **pituitrin** (pi-tu'i-trin), is given in various forms to produce these effects. For example, a type of this extract known as obstetrical pituitrin is given to induce uterine contractions at one of the stages of childbirth.

THE PANCREAS, INSULIN AND DIABETES

Scattered throughout the **pancreas** are small groups of specialized cells called **islets** (i'lets), which are also known as the **islands of Langerhans** (lahng'er-hanz). They function independently, and are *not* connected with the ducts with which the exocrine part of the pancreas is so well supplied. These islands manufacture a hormone known as **insulin**. Insulin is necessary for the normal use of sugar in the body tissues. If for some reason the pancreatic islets fail to produce enough insulin, sugar is not "burned" in the tissues for transformation into energy; instead, the sugar is simply excreted along with the urine. This condition is called **diabetes mellitus** (di-ah-be'teez mel-li'tus).

THE ADRENAL GLANDS AND THEIR HORMONES

The **adrenals**, or **suprarenals**, are two small glands, each one situated above a kidney. An adrenal gland has two separate parts each of which acts as a gland. The inner area is called the **medulla**, while the outer portion is the **cortex**.

The hormones from the medulla

The principal hormone produced by the medulla is one which we already have learned something about: **epinephrine**, also called **adrenaline** (ad-ren'al-in). Another hormone, **norepinephrine** (nor-ep-e-nef'rin), is closely related chemically and is similar but not identical in its actions. These are referred to as the "fight and flight" hormones because of their effects during emergency situations. Some of these effects are:

1. Stimulation of the sympathetic nerves which supply the involuntary muscle in the walls of the arterioles, causing these muscles to contract and the blood pressure to rise accordingly.
2. Conversion of the glycogen of the liver into sugar, which is poured into the blood and brought to the voluntary muscles, permitting them to do an extraordinary amount of work.
3. Increase of the heartbeat rate.
4. Dilation of the bronchioles, through relaxation of the smooth muscle of their walls.

The hormones from the adrenal cortex

This group of hormones has three main functions:

1. To control the reabsorption of sodium in the kidney tubules, and the excretion of potassium. The principal hormone responsible for this electrolyte-regulating function is called **aldosterone** (al-do-ster'one). Recent research suggests that the pineal body produces a hormone called adrenoglomerulotropin (ah-dre-no-glo-mer-u-lo-tro'pin) which activates production of aldosterone by the adrenal cortex.
2. To maintain the carbohydrate reserve of the body by changing amino acids to sugar instead of to protein, whenever the needs of the body call for such action. The hormone that does this is

cortisol (a glucocorticoid). (This hormone is also produced in larger amounts in times of stress and so aids the body in responding to some unfavorable situations.)

3. To govern certain secondary sexual characteristics, particularly in the male. For example, these hormones are responsible in some ways for the muscular vigor that is typically masculine. One of these hormones is sometimes referred to as **adrenosterone** (ad-re-no'ster-one).

It should be mentioned once again that production of these hormones from the adrenal cortex is stimulated by ACTH from the "master gland" (pituitary) which is in turn stimulated by impulses from the hypothalamus. Not only are endocrine glands interrelated so they affect each other, but recent research indicates complex nervous and hormone connections.

The adrenal cortex is essential to life because it is largely by means of this part of the adrenal glands that the body succeeds in adapting itself to the constant changes in the environment.

THE SEX GLANDS

The sex glands, including the ovaries of the female and the testes of the male, are important endocrine structures. The hormones produced by these organs play an important part in the development of the sexual characteristics, usually first appearing in the early teens, and in the maintenance of the reproductive apparatus once full development has been attained. The hormone produced by the male sex glands is called **testosterone** (testos'ter-one), and is responsible for such secondary sexual characteristics as the deep voice and the growth of facial hair, and also for the functioning of certain of the reproductive organs themselves. Those structures which are directly concerned with reproduction are considered the **primary** sexual characteristics.

In the female, the hormone which most nearly parallels testosterone in its action is **estrone** (es'trone), which contributes to the development of the female secondary sexual characteristics, as well as stimulating the development of the mammary glands, the onset of menstruation, and the development and functioning of the female reproductive organs.

There is one other hormone produced by the female sex glands, and it is called **progesterone** (pro-jes'ter-one). This hormone assists in the normal development of the pregnancy; its action will be explained in the next chapter.

Other Hormone-Producing Structures

The **placenta** (plah-sen'tah) produces several hormones during pregnancy (see Chapter 16). These cause changes in the uterine lining, and later in pregnancy they help to prepare the breasts for lactation. The tests for pregnancy are based on the presence of placental secretions.

The **thymus** produces hormones that stimulate production of the small lymphocytes which function in the body's defense against infection. The thymus secretes a substance that has been given the name thymosin (also called thymic hormone). Thymosin promotes the growth of peripheral lymphoid tissue. The thymus is most active during prenatal life and in infancy.

The **pineal** (pin'e-al) body, a small flattened cone-shaped structure located between the two parts of the thalamus, produces a hormone in a number of animals, and possibly also in the human. The hormone, **melatonin** (mel-ah-to'nin), acts on the pigment cells of the skin in certain animals. Melatonin probably regulates release of certain substances from the hypothalamus and these in turn may ultimately regulate secretion of gonadotrophins (hormones that stimulate gonadal activity) from the pituitary (anterior lobe). The human pineal body may be invaded by tumor tissue, and the symptoms produced are probably due to pressure on nearby structures.

The **hypothalamus**, which is located immediately above the pituitary gland, produces substances that act as chemical messengers and they are called neurohormones. They are carried through special blood vessel pathways that extend directly from the hypothalamus to the pituitary. These neurohormones regulate the secretory activity of the pituitary particularly as relates to the stimulation and regulation of the ovaries and testes.

PROSTAGLANDINS

Hormone-like substances called **prostaglandins** (pros-tah-glan'dins) are the object of much research. They were first found in human and animal semen, but it is now known that they exist in most body tissues and fluids. There are at least 16 prostaglandins. It is known that the prostaglandins help to regulate certain organs and structures, including gastrointestinal muscle, uterine muscle and the central nervous system —but just how they produce their effects is not yet known.

Summary

1. **Secretions.**
   A. External: carried from gland to nearby organ or body surface.
   B. Internal: carried to all parts of body by blood or lymph. Called hormones.
2. **Glands:** mostly epithelial, connective tissue framework, fibrous capsule; divided into lobes.
   A. Exocrine: have ducts to carry secretions.
   B. Endocrine: ductless; secretions reach blood via capillaries.
3. **Hormones:** "chemical messengers" regulating activities of various body systems.
4. **Thyroid gland.**
   A. Location and function: in neck. Produces thyroxine; regulates heat and energy generated in tissues; iodine a necessary constituent of thyroxine.
   B. Effects of increased enzyme activity induced by thyroxine include greater use of glucose, fat and protein; less cholesterol and other fatty substances in blood; greater need for vitamins; increased rate of respiration.
5. **Parathyroid glands.**
   A. Location and function: 4 bodies behind thyroid gland. Regulate calcium content of blood.
   B. Coordinate and regulate vitamin D, phosphates and various forms of calcium.
6. **Pituitary gland.**
   A. Location and structure: in skull below brain; two lobes.
      (1) Anterior lobe. Hormones: somatotropic; gonadotropic; thyrotropic; adrenocorticotropic; lactogenic.
      (2) Posterior lobe. Hormone functions: stimulate smooth muscle of blood vessels; reabsorption of water; stimulate uterine contraction.
   B. Neurohormones from hypophysis stimulate pituitary; then pituitary energizes gonads (ovaries and testes).
7. **Pancreas:** islets (or islands) of Langerhans produce insulin.
   A. Function: necessary for proper use of sugar in tissues.
   B. Disorder: diabetes mellitus (caused by insufficient production of insulin).
8. **Adrenal** (or suprarenal) **glands:** 2 of these; each has medulla and cortex.
   A. Adrenal medulla: produces epinephrine (adrenaline). Effects: contraction of arteriole walls; conversion of glycogen to sugar; increased heartbeat; dilation of bronchioles.
   B. Adrenal cortex: hormones control sodium reabsorption, change amino acids to sugar, govern certain secondary sexual characteristics (mostly in male).
   C. Adaptation to changing environment aided by adrenal hormones; necessary for life.
   D. Interrelationships include complex nerve and hormone connections, involving several glands.
9. **Sex glands:** govern development of secondary sexual characteristics and maintain operation of reproductive system.
   A. Testosterone: male hormone.
   B. Estrone and progesterone: female hormones.
10. Placenta, thymus, pineal body and hypothalamus.
11. Hormone-like substances: prostaglandins help regulate gastrointestinal muscle, uterine muscle, central nervous system.

Questions and Problems

1. What is a gland? What is the usual glandular structure? What are the 2 types of glands?
2. What is a secretion? Name the 2 types of secretions.
3. Name some general functions of hormones.
4. Where is the thyroid gland located? What is its hormone and what does it do?
5. How does thyroxine affect enzyme activity?
6. What are some of the effects of enzyme activity on vitamin requirements, heart rate, respiration and muscular and nervous activity?
7. Name and describe briefly 3 tests for thyroid function.
8. What is the main purpose of the parathyroid hormone? What are the effects of removal of these glands? Of excess secretion?
9. Name the 2 divisions of the pituitary and describe the effects of the various hormones of each.
10. Give the abbreviations for and describe some of the anterior pituitary hormones.
11. What is the purpose of insulin in the body? Name and describe the condition characterized by insufficient production of insulin.
12. Name the 2 divisions of the adrenal glands and the effects of the hormones of each.
13. Name the most important glucocorticoid.
14. Which mineralocorticoid is most prevalent?
15. Name the male and female sex hormones and briefly describe what each does.
16. What is the present status in research on the thymus and the pineal body?

CHAPTER SIXTEEN

Specialized reproductive cells •
Male reproductive system •
Female reproductive system •
Pregnancy •
The menopause •

The Reproductive System

This chapter will concern itself with what is certainly one of the most interesting and mysterious attributes of living matter: the ability to reproduce. The lowest forms of life, the one-celled organisms, usually need no partner in order to reproduce; they simply divide by themselves. This form of reproduction is known as **asexual** (nonsexual) reproduction.

In most animals, however, reproduction is **sexual**, meaning that there is a differentiation in the individuals: they are male or female, and both have their own specialized cells designed specifically for the perpetuation of the species. These specialized sex cells are known as **spermatozoa** (sper-mah-to-zo′ah) in the male and **ova** (o′vah) in the female. In spite of the differentiation of the reproductive apparatus in man and woman, both systems have these three broad characteristics in common:

1. Sex glands, or **gonads** (gon′ads or go′nads), which produce the sex cells and manufacture hormones. The sex glands include the **testes**, which are the male gonads, and the **ovaries**, which are the female gonads. Other names for sex cells are **germ cells** and **gametes** (gam′etes).

2. The tubes and passageways necessary for the movement of sex cells.
3. The accessory organs, which include various exocrine glands.

Male Reproductive System

THE TESTES
The male gonads (testes) are normally located outside the body proper in a sac called the **scrotum** (skro′tum), suspended between the thighs. The testes are egg-shaped organs measuring about 1½ to 2 inches (3.7 to 5 centimeters) in length and approximately one inch (2.5 centimeters) in each of the other two dimensions. The bulk of the specialized tissue of the testes is arranged in tubules in the walls of which the spermatozoa are produced. Between these tiny tubes are small groups of cells that secrete the hormone testosterone.

THE MALE TUBES
The tubes for carrying the spermatozoa begin with the tubules inside the testis itself. From these tubes the spermatozoa are collected by a much-coiled tube, 20 feet (600 centimeters or 6 meters) long, called the **epi-**

didymis (ep-e-did'e-mis), which is located inside the scrotal sac. While they are temporarily stored in the epididymis, the spermatozoa mature and become motile, that is, able to move or "swim" by themselves. The epididymis finally extends upward, and then this straighter part becomes the **ductus deferens** (def'er-enz). The ductus deferens continues through a small canal in the abdominal wall and then curves behind the urinary bladder. There each ductus deferens (also called the **vas deferens**) joins with a **seminal vesicle** (ves'e-kal), one on each side of the midline (see Fig. 16.1). The combination of ductus deferens, nerves and blood and lymph vessels extending from the scrotum and testis on each side through the abdominal wall is called the **spermatic cord.**

SEMINAL VESICLES

The seminal vesicles are tortuous muscular tubes with small outpouchings. They are about 3 inches (7.5 centimeters) long and are attached to the connective tissue at the back of the urinary bladder. The glandular lining produces a thick yellow secretion that forms much of the volume of the ejaculated semen. The glucose and other substances in this secretion aid in nourishing the spermatozoa.

THE PROSTATE GLAND

From the ductus deferens and the seminal vesicles the spermatozoa are carried in the vesicle secretion through the **ejaculatory** (e-jak'u-lah-to-re) **ducts**, which are located inside the prostate gland. About 40 tiny tubes from the prostate gland enter these ejaculatory ducts adding the **prostatic** (pros-tat'ik) **secretion** to the sex cells, another secretion that maintains the motility of the spermatozoa. The prostate gland is also supplied with muscular tissue which, upon the order of the nervous system, contracts to aid in the **ejaculation**, the expulsion of the **semen** (the mixture of spermatozoa and secretions) into the urethra and thence to the outside.

THE URETHRA AND THE PENIS

The male urethra, as we saw earlier, serves the dual purpose of conveying urine from the bladder and carrying the reproductive cells and their accompanying secretions to the outside. The ejection of semen into the receiving canal (vagina) of the female is made possible by the **erection**, or stiffening and enlargement, of the penis, through which the longest part of the urethra extends. The penis is made of a spongelike tissue containing many blood spaces that are relatively empty when the organ is flaccid but fill with blood and distend when the penis is erect. The penis and the scrotum are referred to as the external genitalia of the male.

MUCUS-PRODUCING GLANDS

The largest of the mucus-producing glands in the male reproductive system are a pair of pea-sized organs located in the pelvic floor tissues just below the prostate gland. They are called the **bulbourethral** (bul-bo-u-re' thral), or **Cowper's, glands.** Their ducts extend about an inch from each side and then empty into the urethra just before it extends within the penis. Other very small glands secrete mucus into the urethra as it passes through the penis. The mucus from all these glands helps to provide ideal conditions for the spermatozoa to maintain themselves. However, according to some authorities, the prostatic secretion and the mucus serve mainly as lubricants.

THE SPERMATOZOA

The spermatozoa themselves are tiny detached cells. The fact that at least 200 million spermatozoa are contained in the average ejaculation may give some idea of their size. The individual sperm cell is egg-shaped and has a tail which enables it to make its way through the various passages until it reaches the ovum of the female. It is interesting to note that out of the millions of spermatozoa in an ejaculation, only one of these actually fertilizes the ovum. The remainder of the spermatozoa perish within a short time.

THE INGUINAL CANAL

During embryonic life the testis pushes through the muscles and connective tissues of the abdominal wall, carrying with it the blood vessels and the other structures that later form the spermatic cord. This region in the abdominal wall is called the **inguinal** (ing'gwi-nal) **canal.**

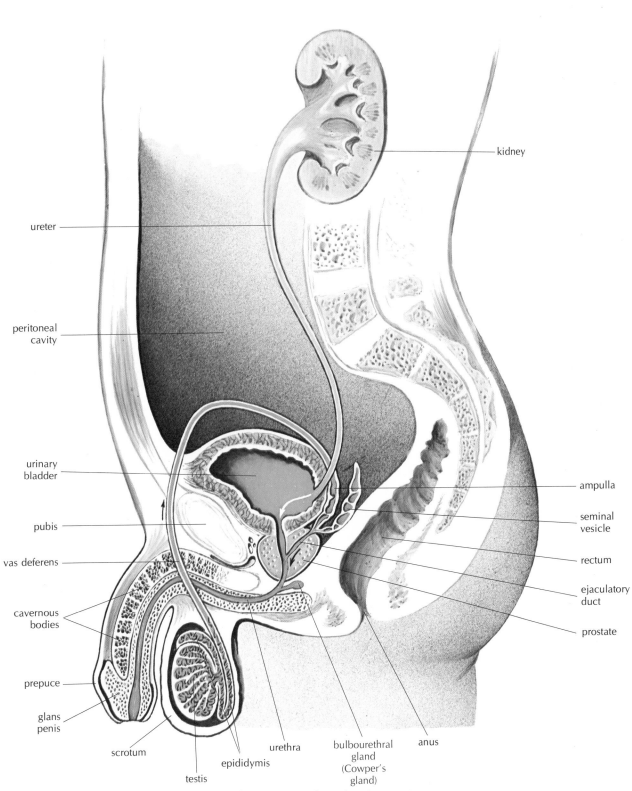

Figure 16.1. Male genitourinary system.

kidney

ureter

peritoneal
cavity

urinary
bladder

pubis

vas deferens

cavernous
bodies

prepuce

glans
penis

scrotum

testis

epididymis

urethra

bulbourethral
gland
(Cowper's
gland)

anus

ampulla

seminal
vesicle

rectum

ejaculatory
duct

prostate

Female Reproductive System

THE OVARIES, OR FEMALE GONADS

In the female the counterparts of the testes are the two **ovaries**, and it is within the ovaries that the female sex cells, or **ova**, are formed. The ovaries are small bodies located in the pelvic part of the abdomen, and are attached to the back of structures made of two layers of peritoneum (called broad ligaments).

The outer layer of each ovary is made of a special epithelium, and it is within this tissue that the ova are produced. The ova are formed and begin a complicated process of maturation in small sacs called **ovarian follicles** (o-va're-an fol'e-kls). When an ovum has "ripened," the ovarian follicle literally ruptures, and the ovum is discharged from the surface of the ovary, making its way to the duct known as the **fallopian** (fah-lo'pe-an) **tube**, one of which exists for each ovary. The rupture of a follicle allowing the escape of the egg cell is called **ovulation** (ov-u-la'shun), and it occurs regularly once a month, probably about halfway between menstrual periods. Although the ovaries produce a vast number of ova, usually only one ovum at a time is released. The regularity of this occurrence is apparently due to the hormones from the pituitary gland.

THE FALLOPIAN TUBES

The egg-carrying tubes of the female reproductive system are variously called **oviducts, uterine** (u'ter-in) **tubes**, or **fallopian tubes**. They are small, muscular structures, nearly 5 inches (12.5 centimeters) long, extending from a point near the ovaries to the uterus (womb). There is no direct connection between the ovaries and these tubes. The ova are swept into the tubes by a current in the peritoneal fluid produced by small fringelike extensions from the edges of the abdominal openings of the tubes. These extensions are called **fimbriae** (fim'bre-e). Once inside the tubes, the ova, which —unlike the spermatozoa—cannot move by themselves, are kept moving toward the uterus by the action of the cilia in the lining of the tubes as well as by peristalsis. It takes about 5 days for the ovum to reach the uterus from the ovary.

THE UTERUS

The organ to which the fallopian tubes lead is the **uterus** (u'ter-us), and it is within this structure that the fetus grows until it is ready to be born.

The uterus is a muscular pear-shaped organ located between the urinary bladder and the rectum. It is approximately 3 inches (7.5 centimeters) long, 2 inches (5 centimeters) wide, and about 1 inch (2.5 centimeters) deep. The upper portion is the larger and is called the body, or **corpus**; the lower, smaller part is the **cervix** (ser'viks), meaning "necklike." The small rounded part above the level of the tubal entrances is known as the **fundus** (fun'dus). The cavity inside the uterus is shaped somewhat like a tiny capital T, but is capable of changing shape and dilating as the embryo (later called the fetus) develops. The cervix leads to the **vagina** (vah-ji'nah), the lower part of the birth canal, which opens to the outside of the body.

The interior layer of the uterus is a specialized epithelium known as **endometrium** (en-do-me'tre-um), and it is this layer which is involved in the phenomenon of menstruation. In order to see the whole picture of this cycle, let us return to the ovaries for a moment.

Remember that the ovum is released about halfway between menstrual periods. While the ovum is still ripening in its follicle, it is surrounded by a fluid which contains the hormone **estrone**. This hormone is carried by the blood to the uterus, where it prepares the endometrium for a possible pregnancy. These preparations include thickening of the endometrium and elongation of the glands that produce the uterine secretion.

The ovum is released, to begin its journey to the uterus. It may or may not become fertilized while passing through the fallopian tube. If it is not fertilized, it disintegrates soon after it reaches the uterus. Also, if fertilization does not occur, the endometrium of the uterus begins to deteriorate. Small hemorrhages appear in this lining, producing the bleeding known as the **menstrual flow**. Bits of the endometrium come away and accompany the menstrual flow. This discharge lasts from 1 to 5 days on the average.

Before the flow ceases, the endometrium begins to repair itself through the growth of new cells from the underlying layer of tissue. Within the ovaries a new ovum is ripening, and the cycle begins anew.

THE VAGINA

The vagina is a muscular tube about 3 inches (7.5 centimeters) long connecting the uterine cavity with the outside. It receives the cervix, which dips into the upper vagina in such a way that a circular recess is formed, giving rise to areas known as **fornices** (for'ne-sez). The deepest of these spaces is behind the cervix and is called the **posterior fornix** (for'niks). This recess in the posterior vagina is separated from the lowest part of the peritoneal cavity by a rather thin layer of tissue, so that abscesses or tumor cells in the peritoneal cavity can sometimes be detected by vaginal examination.

The lining of the vagina is a folded type of mucous membrane something like that found in the stomach. These folds (rugae) permit enlargement so that childbirth will not tear the lining (as a rule). In addition to being a part of the birth canal, the vagina is the organ which receives the penis during sexual intercourse. At or near the vaginal (vaj'i-nal) canal opening to the outside there sometimes may be found a more or less definite fold of membrane called the **hymen**.

THE VULVOVAGINAL GLANDS

Just above and to each side of the vaginal opening are the mucus-producing **vulvovaginal** (vul-vo-vaj'i-nal), or **Bartholin's, glands**. These paired structures are also called the **greater vestibular** (ves-tib'u-lar) **glands** because they open into an area near the vaginal opening known as the **vestibule**. These glands may become infected, then painfully swollen and finally, abscessed. A surgical incision to promote drainage may be required (see Fig. 16.2).

THE VULVA AND THE PERINEUM

The external parts of the female reproductive system form the **vulva** (vul'vah). These include two pairs of lips, or **labia** (la'be-ah), the **clitoris** (kli'to-ris), which is a small organ of great sensitivity, and related structures. Although the entire pelvic floor is properly called the **perineum** in both the male and the female, those who care for the pregnant woman usually refer to the limited area between the vaginal opening and the anus as the perineum (Fig. 16.3).

Pregnancy

FIRST STAGES OF PREGNANCY

The sperm cells are injected by the male into the vagina, and they immediately begin their journey into the uterus and the fallopian tubes. When a sperm cell encounters the ovum, it penetrates the cell membrane of the ovum; the result of the union of these two cells is a cell which can now divide and grow into a new individual. The new cell formed by this union is called the **zygote** (zi'gote). It immediately begins a process of cell division by which a ball of simple cells is formed, and during this time the cilia of the fallopian tube lining are pushing these cells toward the uterine cavity.

As soon as fertilization has taken place, a second female hormone, normally produced to some extent after ovulation has occurred, is manufactured in a greatly increased quantity. To find its origin, let us have another look at the ovary. Once the ovum has left the follicle, the follicle is transformed into a solid glandular mass called the **corpus luteum** (lu'te-um), meaning "yellow body." This yellow body produces its own hormone known as **progesterone** (pro-jes'ter-one), a word meaning roughly "for pregnancy." Normally this hormone takes up where the estrone left off in preparing the endometrium for pregnancy. Now that the ovum has been fertilized, the corpus luteum is much increased in size and its production of progesterone has stepped up accordingly. Since the prepared endometrium is now to be put to use in containing the growing embryo, menstruation ceases. The increased quantity of progesterone brings about the final changes in the endometrium which enable the embryo to be "planted" and nourished within this layer of tissue.

As the pregnancy progresses, the progesterone-forming tissue continues to be stimulated by yet other hormones manufactured by an organ known as the placenta.

After reaching the uterus, the little ball of cells becomes imbedded in the now greatly thickened uterine lining. Then there begins a process of invasion into maternal

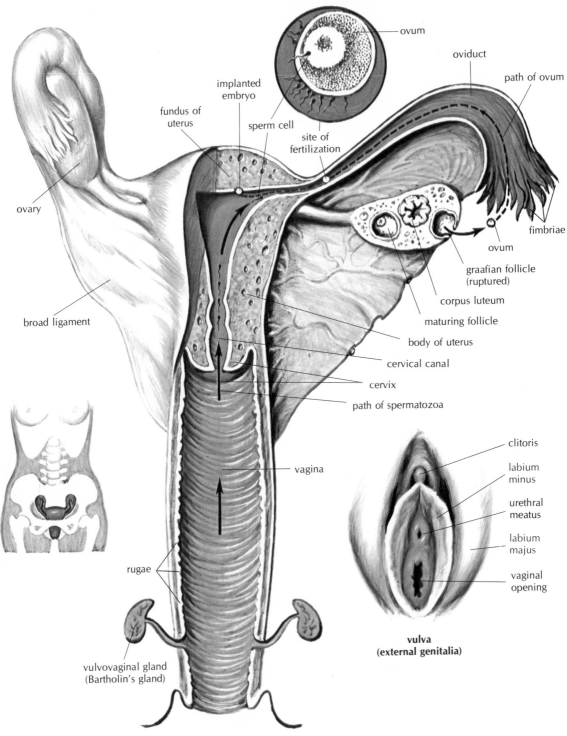

ovum

oviduct

path of ovum

implanted
embryo

sperm cell

fundus of
uterus

site of
fertilization

ovary

broad ligament

fimbriae

ovum

graafian follicle
(ruptured)

corpus luteum

maturing follicle

body of uterus

cervical canal

cervix

path of spermatozoa

vagina

clitoris

labium
minus

urethral
meatus

labium
majus

vaginal
opening

rugae

vulvovaginal gland
(Bartholin's gland)

**vulva
(external genitalia)**

Figure 16.2.   Female reproductive system.

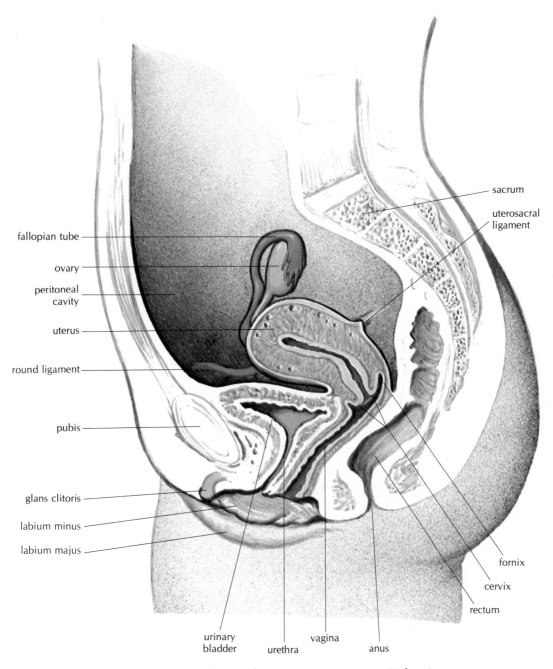

fallopian tube

ovary

peritoneal
cavity

uterus

round ligament

pubis

glans clitoris

labium minus

labium majus

sacrum

uterosacral
ligament

fornix

cervix

rectum

urinary
bladder

urethra

vagina

anus

Figure 16.3. Female reproductive system, as seen in sagittal section.

blood vessels by projections (villi) made of embryonic cells from the ball-like mass. The flat, circular organ called the **placenta** (plah-sen'tah) eventually is formed by this process, serving as the organ of nutrition, respiration, and excretion for the developing individual. The new human being is called an **embryo** (em'bre-o) until the third month;

after that the term **fetus** (fe'tus) is used until birth (Fig. 16.4).

Development of the Embryo
Embryology (em-bre-ol'o-je) is the study of the development of the embryo. This science begins with the fertilization of the ovum and includes a complex series of

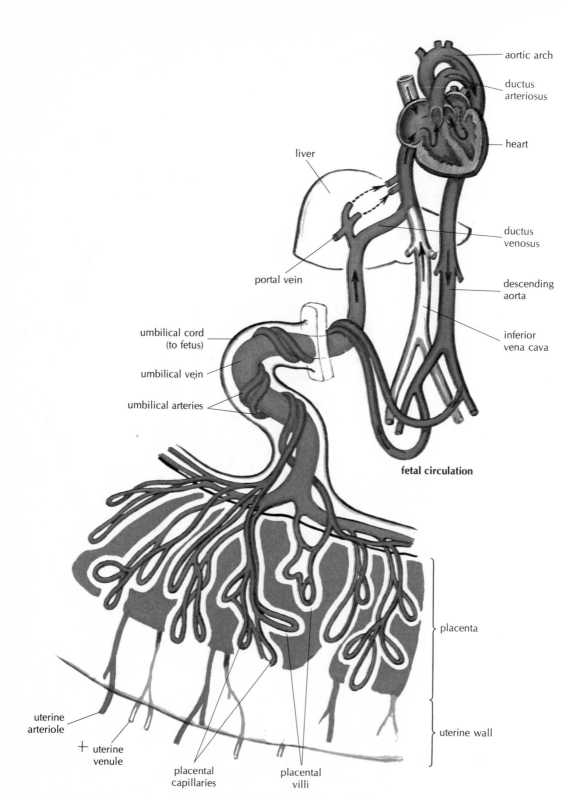

aortic arch

ductus
arteriosus

heart

liver

ductus
venosus

portal vein

descending
aorta

inferior
vena cava

umbilical cord
(to fetus)

umbilical vein

umbilical arteries

fetal circulation

placenta

uterine wall

uterine
arteriole

+ uterine
venule

placental
capillaries

placental
villi

Figure 16.4. Fetal circulation and placenta.

changes about which many books have been written. The embryo develops from only a very small part of the original ball of simple cells. The placenta and the sac that surrounds the embryo as well as the **umbilical** (um-bil'e-kal) **cord** are structures that also originate largely from some of the simple primitive cells of this ball. Among the first organs to develop in the embryo are the heart and the brain. By the end of the first month the embryo is about a fourth of an inch long with four small swellings at the sides called **limb buds**, which will develop into the four extremities. At this time the heart produces a prominent bulge at the front of the embryo. By the end of the second month the embryo takes on an appearance that is recognizably human.

THE FETUS

By the end of the third month the new individual, now known as a fetus, has reached a length of nearly 4 inches (10 centimeters), including the legs. By the seventh month the fetus is usually about 14 inches (35 centimeters) long, while at the end of the pregnancy the normal length is around 18 to 20 inches (35 to 45 centimeters) and the weight varies from 6 to 9 pounds (2.7 to 3.2 kilograms) (Fig. 16.5).

Since the ball of cells became attached to the wall of the uterus, various auxiliary organs designed to serve the fetus have been developing as well. Two of these are the placenta and the umbilical cord, the latter connecting the fetus with the placenta. The cord contains two arteries and a vein. The fetus is encased in a membrane called the **amniotic** (am-ne-ot'ik) **sac**, and this sac is filled with a clear liquid, **amniotic fluid**, which serves as a protective cushion for the fetus. The amniotic sac, which ruptures at birth, is popularly known as the bag of waters. The skin of the fetus is protected by a layer of cheeselike material called **vernix caseosa.**

THE MOTHER

The total period of pregnancy, from fertilization of the ovum to birth, is about 280 days. During this period the mother must supply all the food and oxygen for the fetus, and eliminate its waste materials as well. She must certainly "eat for two," and the

calcium intake should be especially high, since the developing bones and teeth of the fetus call for great quantities of this element. The kidneys of the mother have an especially heavy burden imposed upon them because of the larger amount of nitrogenous waste to be eliminated. For this reason and others it is important that the kidneys function normally.

Nausea and vomiting are present in nearly half of all pregnancies, especially during the early months. These symptoms may be due to a temporary imbalance in body chemistry. There may be a lack of certain hormones or vitamins, a sensitivity to progesterone, or possibly some disturbance of carbohydrate metabolism.

Frequency of urination and constipation are often present during the early stages and then usually disappear as the pregnancy progresses. Late in the pregnancy these symptoms may reappear because the head of the fetus often drops from the abdominal region down into the pelvis where it may press on the rectum and the urinary bladder.

CHILDBIRTH

What triggers the beginning of uterine contractions and the resulting birth of the child? Some possibilities are:

1. A decrease in the placental secretion of progesterone, which normally inhibits uterine contractions.
2. An increase in estrogens produced by the placenta which results in stimulation of the uterine muscles.
3. The production of oxytocin (a natural hormone that stimulates the pregnant uterus) by the posterior pituitary gland, promoting uterine contractions.
4. Mechanical factors such as increase in the size of the fetus plus increase in the movements of the feet and hands that strike the uterine wall and are thought to play a part in causing uterine contractions.
5. Irritation of the cervix by the presenting part, usually the head, which may cause a muscular reaction over the whole of the uterus so that the contractions become strong enough to cause the birth of the child.

The process of giving birth to a child is known medically as **parturition** (par-tu-

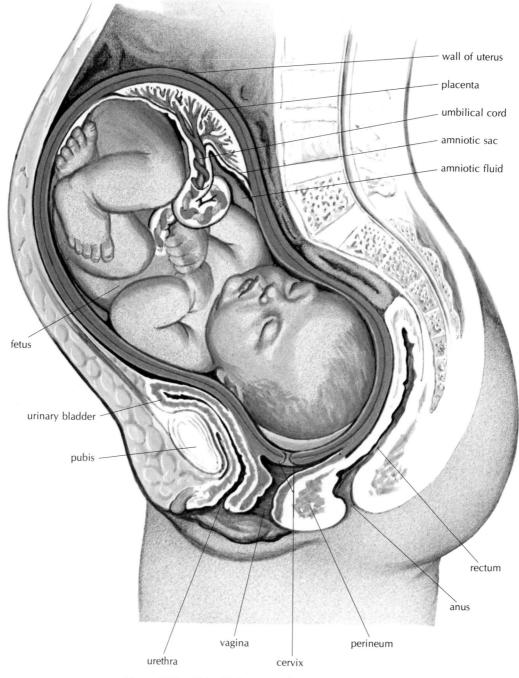

wall of uterus

placenta

umbilical cord

amniotic sac

amniotic fluid

fetus

urinary bladder

pubis

rectum

anus

urethra

vagina

cervix

perineum

Figure 16.5. Midsagittal section of pregnant uterus.

rish'un), and as a rule it is divided into three stages:

1. In the first stage, the muscles of the uterus begin the contractions known as **labor pains**. During this process the amniotic sac is forced into the cervix, serving to dilate it. This stage usually requires several hours, at the end of which time the amniotic sac usually ruptures.

2. This stage involves the passage of the fetus, head first as a rule, through the

cervical canal and the vagina to the outside.

3. During the third stage, usually 15 or 20 minutes after the child is born, the **afterbirth** is expelled. The afterbirth includes the placenta, the membranes of the amniotic sac, and the umbilical cord, except for a small portion remaining attached to the baby's navel, or **umbilicus.**

THE MAMMARY GLANDS AND LACTATION

The **mammary glands**, or the breasts of the female, are accessories of the reproductive system. They are designed to provide nourishment for the baby after its birth; and the secretion of milk at this period is known as **lactation** (lak-ta'shun).

The mammary glands are constructed in much the same manner as the sweat glands. Each of these glands is divided into a number of lobes composed of glandular tissue and fat, and each lobe in turn is subdivided. The secretions from the lobes are conveyed through **lactiferous** (lak-tif'er-us) **ducts**, all of which converge at the nipple (see Fig. 16.6).

The mammary glands begin developing during puberty; they do not become functional until the end of a pregnancy. A lactogenic hormone (**prolactin**), produced by the anterior lobe of the pituitary, stimulates the secretory cells of the mammary glands. The first of the mammary gland secretions is a thin liquid called **colostrum** (ko-los'trum). It is nutritious but has a somewhat different composition from milk. Within a few days milk is secreted, and it will continue for

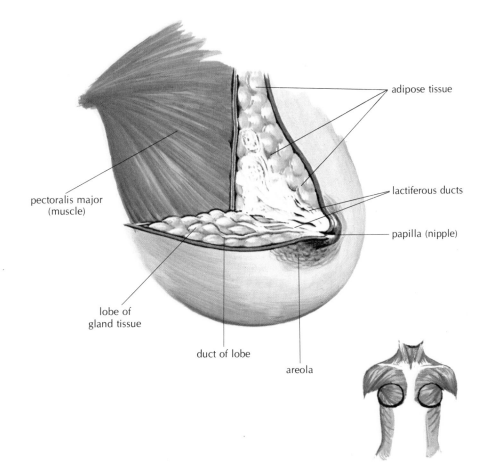

adipose tissue

lactiferous ducts

papilla (nipple)

pectoralis major (muscle)

lobe of gland tissue

duct of lobe

areola

Figure 16.6. Section of breast.

several months if frequently removed by the suckling baby or by pumping.

MULTIPLE BIRTHS

Statistics indicate that twins occur in about 1 in every 80 to 90 births, varying somewhat in different countries. Triplets occur much less frequently, usually once in several thousand births, while quadruplets occur very rarely indeed, and the birth of quintuplets represents an historical event.

Twins originate in two different ways, and on this basis are divided into two groups:

1. **Fraternal** twins are formed as a result of the fertilization of two different ova by two spermatozoa. Two completely different individuals, as different from each other as other brothers and sisters, except in age, are produced. Each fetus has its own placenta and surrounding sac.

2. **Identical** twins develop from a single zygote formed from a single ovum fertilized by a single spermatozoan. Sometime during early stages of development the simple embryonic cells separate into two (or more) groups. Usually there is a single placenta, although there naturally must be an umbilical cord for each individual.

Other multiple births may be fraternal, identical or combinations of these. The tendency to multiple births seems to be hereditary.

Fertility drugs have increased the number of multiple births. These drugs stimulate the ovary either directly or indirectly (via the pituitary).

The Menopause

The **menopause** (men'o-pawz), popularly called the change of life, is that period at which menstruation ceases altogether. It ordinarily occurs between the ages of 42 and 54, the average age being about 48, and is caused by cessation of ovarian activity. The ovary becomes chiefly scar tissue and no longer produces either ova or estrone. Because of the lack of this hormone, the uterus, the fallopian tubes, the vagina and the vulva all gradually become atrophied.

Although the menopause is an entirely normal condition, its onset sometimes brings about effects that are temporarily disturbing. The absence of estrone can cause such nervous symptoms as irritability, "hot flashes" and extreme excitability. Sometimes these symptoms are relieved through the administration of estrone or one of its derivatives. As a rule, the better a woman's state of health, the less traumatic this period will be.

Summary

1. **Reproduction.**
   A. Asexual in some lowest forms of life.
   B. Sexual in most forms. Specialized cells: ova in female, spermatozoa in male.
   C. Reproductive systems in human male and female have in common: gonads, passageways for sex cells, accessory organs.
2. **Male reproductive system.**
   A. Parts and functions: testes (produce spermatozoa and testosterone); tubes (epididymis, ductus deferens); seminal vesicles (form bulk of secretion in semen); prostate gland (contains ejaculatory ducts, produces secretion, muscle produces ejaculation); penis (organ of copulation, contains urethra, erectile tissue); bulbourethral glands (mucus).
3. **Female reproductive system.**
   A. Parts and functions: ovaries (produce ova and estrone; progesterone from corpus luteum); fallopian tubes (convey ova; fertilization occurs here); uterus (lined with endometrium which is prepared by hormones for pregnancy; embryo grows here); vagina (part of birth canal); vulva (labia, clitoris); greater vestibular glands; perineum.
4. **Pregnancy.**
   A. First stages: ovulation; sperm cell fertilizes ovum in fallopian tube; zygote divides, moves to uterus; implanted in uterus; progesterone production increased; menstruation ceases; placenta formed.
   B. Development of embryo: heart and brain

develop early; limb buds after first month; human appearance after second month.

C. Fetus: weighs 6-9 pounds at birth; is in amniotic sac with amniotic fluid; covered with vernix caseosa; umbilical cord connects fetus and placenta.

D. Mother: supplies food and oxygen for fetus; high calcium intake required; kidneys heavily worked.

E. Childbirth. Sequence: labor pains, dilation of cervix, rupture of amniotic sac, passage of the fetus, expulsion of afterbirth.

F. Lactation: secretion of milk by mammary glands (divided into lobes with lactiferous ducts).

G. Multiple births. Twins: fraternal (fertilization of 2 ova by 2 spermatozoa); identical (formed from single zygote).

5. **The Menopause:** final cessation of menstruation, sometimes causing nervous symptoms. These may be relieved by hormone administration.

Questions and Problems

1. In what fundamental respect does reproduction in some single-celled animals such as the ameba differ from that in most animals?
2. Name the sex cells of both the male and the female.
3. Name all the parts of the male reproductive system and describe the function of each.
4. Name the principal parts of the female reproductive system and describe the function of each.
5. Describe the process of ovulation.
6. Describe the processes that occur in menstruation.
7. Name 2 principal female hormones. Where are they produced, and under what circumstances? What does each do?
8. Outline briefly the following processes: conception; development of the embryo and fetus; parturition.
9. Mention some factors that trigger labor and childbirth.
10. Describe each of the 3 stages of childbirth.
11. What differences occur during the early stages of development in fraternal and identical twins?
12. What is the menopause? What can be done to relieve some of its distressing effects?

Suggestions for Further Study

Adrian, R. H., et al.: *Reviews of Physiology,* Vol. 69. New York; Springer-Verlag, 1974.

Anson, B. J., and **McVay, C. B.:** *Surgical Anatomy,* 2 vols., ed. 5. Philadelphia; Saunders, 1971.

Anthony, C. P.: *Basic Concepts in Anatomy and Physiology,* ed. 3. St. Louis; Mosby, 1974.

Anthony, C. P., and **Kolthoff, N. J.:** *Textbook of Anatomy and Physiology,* ed. 8. St. Louis; Mosby, 1971.

Arey, L. B.: *Developmental Anatomy (Embryology).* Philadelphia; Saunders, 1974.

Arey, L. B.: *Human Histology,* ed. 4. Philadelphia; Saunders, 1974.

Bevelander, G.: *Essentials of Histology,* ed. 6. St. Louis; Mosby, 1970.

Brooks, S. M.: *Basic Facts of Body Water and Ions,* ed. 6. New York; Springer, 1973.

Brooks, Stewart: *Basic Science and the Human Body.* St. Louis; Mosby, 1975.

Chaffee, E. E., and **Greisheimer, E. M.:** *Basic Physiology and Anatomy,* ed. 3. Philadelphia; Lippincott, 1974.

Cohen, A., and **Rosen, M. H.:** *Handbook of Microscopic Anatomy for the Health Sciences.* St. Louis; Mosby, 1975.

Crouch, J. E.: *Functional Human Anatomy,* ed. 2. Philadelphia; Lea & Febiger, 1972.

DeCoursey, R. M.: *The Human Organism,* ed. 4. New York; McGraw-Hill, 1974.

Donovan, F.: *Prepare Now for a Metric Future.* New York; Weybright & Talley, 1970.

Dorland's Illustrated Medical Dictionary, ed. 25. Philadelphia; Saunders, 1974.

Elliot, H. C.: *Textbook of Neuroanatomy,* ed. 2. Philadelphia; Lippincott, 1972.

Frances, C. C.: *Introduction to Human Anatomy,* ed. 6. St. Louis; Mosby, 1973.

Ganong, W. F.: *Review of Medical Physiology,* ed. 6. Los Altos, Ca.; Lange, 1973.

Garb, S.: *Laboratory Tests in Common Use,* ed. 5. New York, Springer, 1973.

Gore, R.: "The New Biology, I, The Awesome Worlds within a Cell," National Geographic, Vol. 150, No. 3. Sept. 1976.

Goss, C. M. (ed.): *Gray's Anatomy,* ed. 29. Philadelphia; Lea & Febiger, 1973.

Grant, J. B.: *An Atlas of Anatomy,* ed. 6. Baltimore; William & Wilkins, 1972.

Grant, J. B.: *Grant's Method of Anatomy,* ed. 8. Baltimore; William & Wilkins, 1971.

Greisheimer, E. M., and **Wiedman, M. P.:** *Physiology and Anatomy,* ed. 9. Philadelphia; Lippincott, 1972.

Grollman, S.: *The Human Body,* ed. 3. New York; Macmillan, 1974.

Guyton, A. C.: *Basic Human Physiology,* text ed. Philadelphia; Saunders, 1971.

Guyton, A. C.: *Function of the Human Body,* ed. 4. Philadelphia; Saunders, 1974.

Ham, A. W.: *Histology,* ed. 6. Philadelphia; Lippincott, 1969.

Jacob, S. W., and **Francone, C. H.:** *Structure and Function in Man,* ed. 3. Philadelphia; Saunders, 1974.

Joseph, J.: *Essential Anatomy.* Boston; Herman Publishers, 1974.

Locke, D. M.: *Enzymes, The Agents of Life.* New York; Crown Publishers, 1969.

Lockhart, R. D., et al.: *Anatomy of the Human Body*, text ed. Philadelphia; Lippincott, 1969.

Lopez-Antumez, L.: *Atlas of Human Anatomy*, First United States ed. tr. by H. Monsen. Philadelphia; Saunders, 1971.

Napolitani, F. D.: *Human Anatomy: A Visual Synopsis.* Philadelphia; Lippincott, 1970.

Pansky, B., and **House, E. L.:** *Review of Gross Anatomy*, ed. 2. London, Macmillan, 1969.

Romanes, G. J., ed.: *Cunningham's Textbook of Anatomy*, ed. 11. New York; Oxford University Press, 1971.

Schottelius, B. A., and **Schottelius, D. D.:** *Textbook of Physiology*, ed. 17. St. Louis; Mosby, 1973.

Shepro, D.: *Human Anatomy and Physiology.* New York; Holt, 1974.

Sobotta, J.: *Atlas of Human Anatomy*, 3 vols., ed. 9. New York; Macmillan, 1974.

Thomas, V. E., and **Mansir, A. R.:** *Life Sciences for Health Technologies.* Los Angeles; Technicourse, Inc., 1970.

Warwick, R., and **Williams, P. L.,** eds.: *Gray's Anatomy*, ed. 35. Philadelphia; Saunders, 1973.

Glossary

abdomen (ab-do′men): the part of the body located between the diaphragm and the pelvis; the cavity that contains the abdominal organs (viscera). It is also called the **venter**.

abducens (ab-du′senz): a nerve which sends controlling impulses to a muscle of the eyeball.

abduction (ab-duk′shun): a movement away from the axis or midline of the body; the act of turning outward.

accommodation (ah-kom-o-da′shun): a change in the shape of the eye lens so that vision is more acute; an adjustment of the eye lens for various distances; the focusing process.

acetone (as′e-tone): an organic compound that may be found in abnormal amounts in the urine of diabetics. Acetone bodies are also called ketone bodies.

acoustic (ah′koos-tik): pertaining to sound or the sense of hearing.

acromion (ah-kro′me-on): the flattened projection that extends from the shoulder blade (scapula) to form the top of the shoulder itself; also called the **acromial process**.

adduction (ah-duk′shun): a movement toward the body, or toward the midline of the body; the act of turning inward.

adipose (ad′e-pose): fat; containing fat, as adipose tissue.

adrenal (ad-re′nal): located near the kidney; specifically the endocrine gland near the kidney. It is also called the **suprarenal gland**.

adrenocorticotropic hormone (ah-dre-no-kor-ti-ko-trop′ik): a hormone which stimulates the adrenal cortex. Abbreviated ACTH.

adrenoglomerulotropin (ah-dre-no-glo-mer-u-lo-tro′pin): a hormone, probably produced by the pineal body, that activates the production of aldosterone by the adrenal cortex.

adrenotropic (ad-ren-o-trop′ik): having an influence over the adrenal (suprarenal) glands.

afferent (af′er-ent): carrying toward a center or main part; nerves that carry impulses toward the central nervous system, or toward ganglia.

agglutination (ah-gloo-ti-na′shun): a process by which cells (bacteria, blood cells, etc.) collect in groups or clumps; clumping.

albumin (al-bu′min): a body protein found in urine when there is kidney damage.

aldosterone (al-do-ster′on): a hormone produced by the adrenal cortex which regulates electrolytes.

alimentary canal (al-e-men′ter-e): a continuous passageway from the mouth, where food is taken in, to the anus, where waste products are discharged; the process of digestion takes place here.

allergen (al′er-jen): a substance that causes sensitivity; something that induces allergy.

alveoli (al-ve′o-li): a cluster of air sacs at the end of the bronchial trees; sockets for the teeth or any small hollows or cavities. The singular form is **alveolus**.

ameba (ah-me′bah): a single-celled protozoan, a minute irregular mass of protoplasm which propels itself by extending a branch, or "false foot," and then flowing over it.

ameboid (ah-me′boid): resembling an ameba in appearance, or more especially in movement (by means of protoplasmic flowing and "false feet").

amino acids (am-e′no): the building blocks of protein.

amniocentesis (am-ne-o-sen-te′sis): the perforation of the abdominal wall and the uterine wall in order to remove amniotic fluid for examination.

amniotic fluid (am-ne-ot′ik): a clear liquid which serves as the protective cushion for the fetus.

anaphylaxis (an-ah-fi-lak′sis): a severe reaction caused by extreme sensitivity to a foreign protein or other substance. The adjective is **anaphylactic** (an-ah-fi-lak′tik).

anastomosis (ah-nas-to-mo′sis): a connection between blood vessels; a surgical formation of a passage between 2 distinct spaces or 2 organs (or their parts).

anatomy (ah-nat′o-me): the science of the structure of the body and the relationship of its parts to each other.

anemia (ah-ne′me-ah): a decrease in certain elements of the blood, especially red cells and hemoglobin.

anteflexion (an-te-flek′shun): an abnormal forward curvature; a bending forward of the upper part of an organ, usually abnormal, but regarded as normal for the uterus.

anterolateral (an-ter-o-lat′er-al): located in front of and to one side.

anus (a′nus): the outside opening of the anal canal.

aorta (a-or′tah): the largest artery.

apnea (ap′ne-ah): a temporary cessation of breathing.

aponeurosis (ap-o-nu-ro′sis): a sheetlike layer of connective tissue connecting a muscle to the part that it moves, or acting as a sheath enclosing a muscle.

appendicular (ap-en-dik′u-lar): that part of the skeleton which forms the framework for the extremities.

arrhythmia (ah-rith′me-ah): a lack of normal rhythm, especially of the heart beat.

arteriole (ar-te′re-ole): the smallest artery, one that branches into the microscopic capillaries.

atom (at′om): any one of the ultimate units of an element that can exist and still have the properties of the element; the particles that together form a molecule in a compound.

audiovisual (aw-de-o-vizh′u-al): stimulating the senses of both hearing and sight (said of aids such as slides and motion pictures used in teaching).

auricle (aw′re-kl): the projecting part of the ear.

automaticity (au-to-mah-tis′e-te): a characteristic whereby there is action without an outside stimulus. An example is the contraction of the heart muscle.

autonomic (aw-to-nom′ik): pertaining to the division of the nervous system which controls more or less automatic activities.

axial (ak′se-al): that part of the skeleton which includes the bony framework of the head and trunk.

axilla (ak-sil′ah): the small hollow beneath the arm where it joins the body at the shoulder; the armpit.

basal ganglia (ba′sal gang′le-ah), **basal nuclei**: masses of gray matter within the lower part of the forebrain which aid in maintaining muscle coordination and steadiness of muscle contraction.

bifurcate (bi-fur′kate): to divide into 2; forming 2 branches or subdivisions.

biliary (bil′e-a-re): relating to bile, the gallbladder or the bile ducts.

biopsy (bi′op-se): removal of tissue or other material from the living body for purposes of examination. It is usually a microscopic study.

brachial (bra′ke-al): pertaining to the arm (the part between the shoulder and the elbow).

brachiocephalic (brak-e-o-se-fal′ik): relating to the arm and the head, as for example, the brachiocephalic artery.

bronchiole (brong′ke-ole): a very small subdivision of the lung tubes; a microscopic bronchial tube; also called a **bronchiolus** (brong-ki′o-lus).

bronchus (brong′kus): a main division of the trachea (windpipe); one of the larger air passages in the lungs. The plural form is **bronchi** (brong′ki).

calorie (kal′o-re): a unit of heat. (The small calorie is the amount of heat required to raise the temperature of 1 gram of water

1 degree centigrade. The large calorie or kilocalorie is the one used in nutrition and metabolic studies, and is the amount of heat necessary to raise 1 kilogram of water 1 degree centigrade).

calyx, calix (ka′liks): an extension from the renal pelvis into the kidney tissue. The plural is either **calyces** or **calices** (kal′i-sez).

carbohydrate (kar-bo-hi′drate): a starch, sugar, cellulose or gum; a compound containing carbon, hydrogen and oxygen in a particular amount and arrangement.

cardiac (kar-de-ak′): pertaining to the heart.

carotid (kah-rot′id): relating to the principal artery extending up through the neck to the head.

cecum (se′kum): a small pouch at the beginning of the large intestine.

celiac (se′le-ak): relating to or located in the abdomen.

Celsius (sel′se-us): the scale generally used to measure temperature. Freezing point is 0 degrees and boiling point is 100 degrees. Also known as the **centigrade scale**.

centigrade (sen′ti-grad): a unit of measurement consisting of 100 steps or degrees of temperature. Also called the Celsius scale.

centimeter (sen′ti-me-ter): a unit of measurement, being one hundredth part of a meter.

centrifugal (sen-trif′u-gal): moving away from a center.

centrifuge (sen′tri-fuj): a machine by which tubes of a mixture, solution or suspension are rapidly revolved so that floating particles are driven away from the center of rotation. An example is the separation of the plasma from the formed elements of the blood.

centrioles (sen′tri-ols): the 2 bodies of the centrosome which separate, during mitosis, initiating the formation of 2 daughter cells.

centrosome (sen′tro-som): a specialized area in the cell protoplasm just outside the nucleus which plays an essential part in cell division.

cerebellum (ser-e-bel′um): the part of the hindbrain that lies below the occipital part of the cerebrum on each side, concerned with voluntary muscle movement.

cerebrospinal fluid (ser-e-bro-spi′nal): fluid formed in the ventricles of the brain which cushions the organ against shock.

cerebrovascular (ser-e-bro-vas′ku-lar): pertaining to the blood vessels of the brain.

cerebrum (ser′e-brum): the largest part of the brain located in the upper portion of the cranium, consisting of 2 cerebral hemispheres which are divided into lobes.

cerumen (se-roo′men): earwax.

cervical (ser′vi-kal): relating to the neck or any cervix, including the cervix of the uterus.

cervix (ser′viks): any neck or constricted portion of an organ, part or region of the body.

cholesterol (ko-les′ter-ol): an organic fat-like compound, found in animal fat, bile, blood tissue, the nerve fiber sheath (myelin), liver and other parts of the body.

choroid (ko′roid): pertaining to the thin, dark brown, vascular middle coat of the eyeball; also relating to the capillary fringe-like parts of the pia mater that extend into the brain ventricles and produce cerebrospinal fluid.

chromosomes (kro′mo-somes): small rod-shaped bodies that stain deeply and appear in the nucleus at the time of cell division (mitosis). They contain the hereditary factors, the genes.

chyle (kile): the milky fluid in the lacteals of the small intestine after digestion; a combination of lymph and emulsified fat.

chyme (kime): the semifluid creamy or gruel-like material in the stomach; the material formed by the mixing and partial digestion of food in the stomach.

cilia (sil′e-ah): hairs or hairlike processes (may refer to eyelashes or to microscopic extensions of the cell protoplasm). The singular form is **cilium** (sil′e-um).

circumcision (ser-kum-sizh′un): removal of the foreskin, the fold over the glans penis in the male.

cisterna chyli (sis-ter′nah ki′li): the temporary storage pouch for chyle and lymph before they are carried by the thoracic duct into the blood stream.

conchae (kong′ke): shell-shaped structures. The singular form is **concha** (kong′kah).

congenital (kon-jen′i-tal): present at and usually before birth.

conjunctiva (kon-junk-ti′vah): the thin delicate membrane that lines the eyelids and is reflected over the front of the eyeball.

convolution (kon-vo-lu′shun): an elevation caused by an infolding of a structure upon itself.

corium (ko're-um): the deeper part of the skin found below the epidermis; the dermis or true skin, containing blood vessels, nerves and connective tissue.

cornea (kor'ne-ah): the transparent front part of the eyeball; the forward continuation of the outer coat (sclera).

coronary (kor'o-na-re): applying to structures that encircle a part or organ in a crownlike manner, as for example, the coronary arteries encircling the base of the heart.

corpus (kor'pus): the main part of a structure or organ. It may designate the whole organism.

corpuscle (kor'pus'l): a living cell; a small mass or body.

corpus luteum (kor'pus lu'te-um): a yellow mass formed in the ovarian follicle which produces the hormone progesterone.

creatinine (kre-at'i-nin): a nitrogen waste product found in urine.

cytology (si-tol'o-je): the study of cells.

cytoplasm (si'to-plazm): the cell protoplasm outside the nucleus.

deciduous (de-sid'u-us): relating to anything that is cast off at maturity, as for example, the first set of teeth.

defecation (def-e-ka'shun): the discharge of fecal material from the rectum.

degeneration (de-jen-er-a'shun): breaking down, as in deterioration of tissue with aging.

dehydration (de-hi-dra'shun): a condition due to excessive water loss from the body or its parts.

dendrite (den'drite): a nerve fiber which conducts impulses to the cell body.

deoxyribonucleic (de-ok-se-ri-bo-nu-kle'ik) **acid**: the nucleic acid primarily found in the nucleus of cells; the hereditary material of each cell. Also called **DNA**.

dialysis (di-al'i-sis): the process of separating crystalloids (smaller particles) from colloids (larger particles) by the difference in their rates of diffusion through a semipermeable membrane.

diaphragm (di'ah-fram): any partition that separates one area from another, especially the somewhat dome-shaped musculomembranous partition between the thoracic and abdominal cavities.

diaphysis (di-af'i-sis): the shaft of a long bone.

diastole (di-as'to-le): the relaxing dilatation period of the heart muscle, especially of the ventricles. The adjective is **diastolic** (di-ah-stol'ik).

diffusion (di-fu'shun): the process of becoming spread out, as any gaseous substance that spreads throughout a room; the movement of molecules from any area of high concentration to one of lower concentration.

dilation (di-la'shun): enlargement of the pupil of the eye. May also apply to the stretching or enlargement of other parts of the body, such as the cervix.

distal (dis'tal): farther from a point of reference. The opposite is proximal.

dominant (dom'i-nant): in genetics, means capable of expression in the offspring even though only one of a pair of opposite Mendelian characters is present.

duodenum (du-o-de'num): the first or proximal part of the small bowel (about 12 inches in length).

dura mater (du'rah ma'ter): the outermost, toughest of the 3 coverings of the brain and spinal cord.

dyspnea (disp'ne-ah): difficult or labored breathing.

echocardiography (ek'o-kar-de-og-rah-fe): a method for detecting abnormalities of the heart using high-frequency sound impulses.

efferent (ef'er-ent): carrying away from a center.

effusion (e-fu'shun): the escape of fluid into a space or part; the fluid that has escaped.

electrocardiogram (e-lek-tro-kar'de-o-gram): the tracing of the electric current produced by heart muscle activity (contraction); the record produced by an electrocardiograph.

electrocardiograph (e-lek-tro-kar'de-o-graf): an instrument used for making records of the heart's electric currents.

electroencephalogram (e-lek-tro-en-sef'ah-lo-gram): the tracing of the electric currents developed in the brain; the record produced by the **electroencephalograph**.

electroencephalograph (e-lek-tro-en-sef'ah-lo-graf): an instrument for making records of the brain's electric currents.

electrolyte (e-lek'tro-lite): a solution that conducts electricity by means of ions that are positively or negatively charged.

electron (e-lek'tron): the unit of negative electricity.

element (el'e-ment): in chemistry, a simple substance that cannot be decomposed into simpler substances by chemical means; any one of the primary parts of a thing.

embryology (em-bre-ol'o-je): the study of the development of the embryo.

endocardium (en-do-kar'de-um): the membrane which lines the heart chambers and assists in forming the heart valves.

endocrine (en'do-krin): secreting to the inside, into either tissue fluid or blood.

endometrium (en-do-me'tre-um): the special layer of epithelium which forms the inner lining of the uterus.

endothelium (en-do-the'le-um): the layer of cells that lines blood and lymph vessels, the heart and the serous body cavities.

epicardium (ep-i-kar'de-um): the membrane that forms the outer layer of the heart wall and is continuous with the lining of the sac that encloses the heart; the visceral pericardium.

epidermis (ep-i-der'mis): the outer epithelial layer of the skin, which contains no blood vessels and which rests on the dermis.

epigastrium (ep-i-gas'tre-um): the upper middle section of the abdominal cavity, just below the breastbone.

epiglottis (ep-i-glot'is): a lidlike structure made largely of cartilage, which covers the entrance to the larynx.

epinephrine (ep-i-nef'rin): a hormone produced by the adrenal medulla; also produced synthetically.

epiphysis (e-pif'i-sis): the end of a long bone, usually larger in diameter than the shaft.

epithelium (ep-i-the'le-um): the tissue that forms the outer part (epidermis) of the skin; it lines blood vessels, hollow organs and passages that lead to the outside of the body, and makes up the active part of many glands.

erythroblastosis (e-rith-ro-blas-to'sis): the presence of immature red cells (erythroblasts) in the circulating blood. Erythroblastosis fetalis: a congenital disorder in which an Rh negative mother transmits antibodies against the Rh protein to an Rh positive baby.

erythrocyte (e-rith'ro-site): a red blood cell.

esophageal (e-sof-ah-je'al or e-so-fa'je-al): relating to the gullet. See **esophagus**.

esophagus (e-sof'ah-gus): the gullet (the tubular passage extending from the pharynx to the stomach).

etiology (e-te-ol'o-je): the study of causes of disease, including theories of origin and organisms that may be involved in causation.

eustachian (u-sta'ke-an) **tube**: a tubelike structure connecting the middle ear cavity and the throat which equalizes air pressure.

excretion (eks-kre'shun): discharging or throwing off waste matter.

exocrine (ek'so-krin): secreting toward the outside, or away from the secreting tissue by tubes or ducts.

exophthalmos (ek-sof-thal'mos): an abnormal bulging of the eyeball.

extracellular (eks-trah-sel'u-lar): outside a cell or cells.

fallopian tubes (fah-lo'pe-an): ducts connected to the uterus. They carry ova from the vicinity of the ovaries into the uterine cavity and provide a place for the fertilization of the ovum by the spermatozoon.

fascia (fash'e-ah): a layer or band of connective tissue, especially that which holds the skin to the surface muscles, or any of the sheets enclosing muscles or other organs.

feces (fe'sez): the material discharged from the bowel which is made up of bacteria, secretions and food residue, also called stool.

fetal (fe'tal): pertaining to the developing unborn baby after the first 8 weeks. See **fetus**.

fetus (fe'tus): the unborn offspring of the human after 8 weeks. (Before 8 weeks it is called an embryo.)

fibrinogen (fi-brin'o-jen): a protein found in the liquid (plasma) of the blood which is converted into insoluble fibrin during clotting.

filtration (fil-tra'shun): the passage of a liquid through a filter or a membrane that acts as a filter.

fimbriated (fim'bre-at-ed): fringed, as the lateral ends of the uterine tubes (fallopian tubes or oviducts).

flaccid (flak'sid): weak, lax and soft.

flatus (fla'tus): gas, usually air, in the stomach or bowel.

flexion (flek'shun): a bending movement of a part of the body.

fluoroscope (floo-o'ro-skope): an instrument used to examine deep structures of the body by means of roentgen rays (x-rays).

fontanel, fontanelle (fon-tah-nel'): a soft spot in a baby's skull; a membrane-covered space where ossification (bone formation) has not yet occurred.

foramen (fo-ra'men): a natural opening or passageway; a general term especially for a passage into or through a bone. The plural is **foramina** (fo-ram'i-nah).

fossa (fos'sah): a hollow or depressed area; a valleylike region on a bone or other structure. The plural form is **fossae**.

fovea (fo've-ah): a small pit or cup-shaped depression in the surface of a part or organ, as in the head of the femur and near the center of the retina of the eye; the point of clearest vision.

gamma globulins (gam'mah glob'u-lins): protein substances often found in immune serums that act as antibodies.

ganglion (gang'gle-on): a small knotlike mass; in the nervous system, a collection of nerve cells.

gene (jene): one of the biologic units of heredity. Genes are parts of the DNA molecule and each is located in a definite position on a certain chromosome.

genetics (je-net'iks): the study of heredity.

genitourinary (jen-i-to-u'ri-nar-e): relating to the organs of both the urinary and the reproductive systems.

gingiva (jin'ji-vah): the mucous membrane and connective tissue that encircles the neck of the tooth and overlies the crown of those not yet erupted. Infection of the gum is called **gingivitis** (jin-ji-vi'tis).

glomerulus (glo-mer'u-lus): a ball-like cluster of nerves or blood vessels, especially the microscopic tuft of capillaries that is surrounded by the expanded part of each kidney tubule.

glossopharyngeal (glos-o-fah-rin'je-al): pertaining to the tongue and the pharynx.

glycogen (gli'ko-jen): the so-called animal starch, the chief storage carbohydrate in animals. It is stored largely in the liver and is liberated in the form of glucose as needed by the body cells.

gonad (gon'ad): an ovary or a testis; a sex gland.

gustatory (gus'tah-to-re): pertaining to the sense of taste.

gynecology (gi-ne-kol'o-je): the branch of medicine related to the study and treatment of disorders of the female reproductive system.

hematocrit (he-mat'o-krit): the volume percentage of red blood cells in whole blood.

hemocytometer (hem-o-si-tom'e-ter): an instrument used for counting blood cells.

hemoglobin (he-mo-glo'bin): the oxygen-carrying colored compound in the red blood cells.

hemoglobinometer (he-mo-glo-bi-nom'e-ter): an instrument used for measuring the amount of hemoglobin in the blood. It is also called a **hemometer** (he-mom'e-ter).

hemolysis (he-mol'i-sis): the disintegration of red blood cells which results in the appearance of hemoglobin in the surrounding fluid.

hemolytic (he-mo-lit'ik): having the ability to disintegrate or hemolyze (he'mo-lize) red blood cells. See **hemolysis**.

hemophilia (he-mo-fil'e-ah): a hereditary blood disorder in which there is deficient production of certain factors involved in blood clotting, resulting in bleeding into joints and deep tissues.

heparin (hep'ah-rin): a complex acid compound, found most abundantly in the liver, which prevents blood clotting; in pharmacy, a mixture obtained from animal livers or lungs which is used in the prevention and treatment of disorders involving blood coagulation.

heredity (he-red'i-te): the transmission of certain characteristics from parents to their offspring, via the genes.

hilum (hi'lum), **hilus** (hi'lus): an area, depression or pit where blood vessels and nerves enter or leave the organ.

histamine (his'tah-min): a substance released from the tissues during the antigen-antibody reaction, which may cause allergic responses.

homeostasis (ho-me-o-sta'sis): a consistency and uniformity of the internal body environment which maintains normal body function; stability of body fluids and their constituents.

hypochondriac (hi-po-kon'dre-ak): pertaining to the region below the costal cartilages at each side of the upper abdominal cavity.

hypogastric (hi-po-gas'trik): pertaining to the lower central region of the abdominal cavity below the umbilical region.

hypothalamus (hi-po-thal'ah-mus): a part of the forebrain near the third ventricle, containing groups of nerve cells that control temperature, sleep, water balance and other chemical and visceral activities.

ileum (il'e-um): the last or distal part of the small intestine, ending at the cecum of the large intestine.

ileus (il'e-us): obstruction of the small bowel which may be due to paralysis or to a mechanical blockage.

iliac (il'e-ak): pertaining to the bone called the ilium (upper os coxae); the 2 regions at each side of the lower abdomen.

ilium (il'e-um): the upper wingshaped portion of the os coxae (hip bone).

inguinal (in'gwi-nal): relating to the groin region.

inheritance (in-her'i-tans): those characteristics that are transmitted from parents to their offspring.

integument (in-teg'u-ment): a covering, especially the skin.

integumentary (in-teg-u-men'tar-e): composed of or relating to the skin; serving as a covering.

intercostal (in-ter-kos'tal): situated between the ribs (costae).

interstitial (in-ter-stish'al): pertaining to spaces or structures between the functioning active tissues of any part or organ.

intracellular (in-trah-sel'u-lar): within the cell itself.

islands of Langerhans (lang'er-hanz): groups of specialized cells scattered throughout the pancreas; one type, the beta cells, secrete insulin, and degeneration of these cells is one of the causes of diabetes.

jejunum (je-joo'num): the second portion of the small intestine; the part of the small intestine between the duodenum and the ileum.

jugular (jug'u-lar): pertaining to the veins of the neck which drain the areas supplied by the carotid arteries.

karyotype (kar'e-o-tip): the arrangement of chromosomes that is characteristic of the species or of a certain individual.

lacrimation (lak-re-ma'shun): the secretion and discharge of tears.

lactation (lak-ta'shun): the secretion of milk.

lacteal (lak'te-al): related to milk; one of the intestinal lymph vessels which take up fat from digested food.

larynx (lar'inks): the voice box.

laser (la'zer): a device that produces a very highly concentrated and intense beam of light.

leukocyte (lu'ko-site): a white blood cell; any colorless ameboid cell mass.

locus (lo'kus): the specific location or site of a gene within the chromosome.

lumen (lu'men): the channel or space inside a tube or a tubular organ.

lymph (limf): a yellowish relatively clear watery fluid found in the lymphatic vessels; a liquid containing cells, mostly lymphocytes, and after a meal, fat globules; any clear watery fluid resembling true lymph.

lymphatic (lim-fat'ik): relating to the system of vessels that contain lymph.

lymphocyte (lim'fo-site): a white blood cell (leukocyte) which is formed in lymphoid tissue including the spleen, thymus, and lymph nodes.

macroblast (mak'ro-blast): a very large nucleated red blood cell. It is also called a **megaloblast.**

medulla (me-dul'lah): the innermost part of an organ, as seen in the kidneys and adrenal glands; the part of the brain that connects with the spinal cord.

megakaryocyte (meg-ah-kar'e-o-site): a giant cell found in the bone marrow; a cell with a large irregularly shaped (or lobulated) nucleus, believed to give rise to the blood platelets.

meiosis (mi-o'sis): a special method of cell division occurring during the development of sex cells (ova and sperm) in which the number of chromosomes is reduced, so that there are only half as many in the mature gamete as there are in other body cells of the species.

melanin (mel'ah-nin): the dark pigment found in some parts of the body, such as the skin, the middle coat of the eye and certain tissues in the brain.

melatonin (mel-ah-to'nin): a hormone which acts on the pigment cells of the skin.

Mendelian laws (men-de'le-an): principles of heredity discovered by an Austrian monk named Gregor Mendel, including the law of independent characteristics, such as height, color, etc., that are inherited as separate units.

meninges (me-nin'jez): the 3 membranes that cover the brain and spinal cord.

menopause (men'o-pawz): the cessation of the normal monthly uterine bleeding.

menses (men'sez): the monthly flow of blood from the female genital tract.

mesentery (mes'en-ter-y): the membranous peritoneal fold which attaches the small intestine to the dorsal body wall.

metabolism (me-tab'o-lizm): the physical and chemical changes or processes by which living substance is maintained and by which energy is produced.

metacarpus (met-ah-kar'pus): the part of the hand near the wrist, between the wrist and fingers; the 5 elongated bones in the hand.

micturition (mik-tu-rish'un): the act of expelling urine from the bladder; urination.

mitochondria (mit-o-kon'dre-ah): very small rod-shaped structures or granules in the cytoplasm of cells which are responsible for oxidative reactions to release energy from the food materials.

mitosis (mi-to'sis): indirect cell division by which the two daughter cells receive identical complements of chromosomes characteristic of the somatic cells of the species.

molecule (mol'e-kule): a minute mass of matter; a combination of atoms that form a given chemical substance or compound; the smallest particle in a chemical compound that can exist in a free state.

mucosa (mu-ko'sah): a lining membrane that produces mucus and is found in spaces connected with the outside, such as the alimentary and respiratory tract; mucous membrane.

mutation (mu-ta'shun): a variation in an inheritable characteristic, a permanent transmissible change in which the offspring differ from the parents.

myelin (mi'e-lin): the fatlike substance that forms a covering for many nerve fibers. These are called myelinated nerve fibers.

myelogenous (mi-e-loj'e-nus): relating to or produced in the bone marrow.

myocardium (mi-o-kar'de-um): the middle thick layer of the heart wall. It is composed of cardiac muscle.

myometrium (mi-o-me'tre-um): the middle muscular wall forming the bulk of the uterus.

myoneural (mi-o-nu'ral): pertaining to the motor nerve fibers.

nephron (nef'ron): the microscopic functional unit of kidney tissue, consisting of a glomerulus with its capsule, convoluted tubules, and Henle's loop, plus the collecting tubule.

neurilemma (nu-re-lem'mah), **neurolemma** (nu-ro-lem'mah): a very thin membrane wrapping the nerve fibers of the peripheral nervous system.

neuroglia (nu-rog'le-ah): the special connective tissue of the central nervous system.

neuron (nu'ron): the nerve cell body plus its processes; the structural unit of nerve tissue.

norepinephrine (nor-ep-e-nef'rin): one of the "fight or flight" hormones; related to epinephrine, both are put to use in emergency situations.

nucleolus (nu-kle'o-lus): a tiny globule located within the nucleus of a cell.

nucleus (nu'kle-us): a small spherical body within a cell; a general term used to indicate a group of nerve cells, usually connected with the fibers of a particular nerve; in chemistry, the central part of an atom.

obligate (ob'li-gate): compulsory or necessary.

occipital (ok-sip'i-tal): relating to the back part of the head (the occiput).

occlusion (o-kloo'zhun): the process of closing or the state of being closed; also applied to the manner of bringing the upper and lower teeth together.

oculomotor (ok-u-lo-mo'tor): relating to eye movements, as the oculomotor nerve supplying many of the eye muscles.

olfactory (ol-fak'to-re): pertaining to the sense of smell.

ophthalmic (of-thal'mik): relating to the eye, as the ophthalmic arteries, veins, and nerves.

organelle (or-gah-nel'): a specific particle of living material present in most cells and serving a specific function in the cell.

organism (or'gan-ism): an individual animal or plant; any organized living thing.

osmosis (os-mo'sis): the passage of a pure solvent, such as water, from a solution of lesser concentration to one of greater concentration through a semipermeable membrane.

osseous (os'e-us): bony; resembling or having the quality of bone.

ossicle (os'sik-l): a small bone, such as those within the middle ear.

ossification (os-i-fi-ka'shun): the process of forming bone, or the conversion of fibrous tissue, or cartilage, into bone.

ova (o'vah): the female reproductive cells; eggs. The singular form is **ovum**.

ovulation (ov-u-la'shun): the discharge of a mature egg cell (ovum) from the follicle of the ovary.

pancreas (pan'kre-as): a large, elongated gland behind the stomach which aids in digestion.

Papanicolaou's (pap-ah-nik-o-la'ooz) **stain**: cells from the upper vagina are stained and studied for the presence of cancer.

parathyroid (par-ah-thi'roid): located near the thyroid gland in the neck; any of the 4 small glands embedded in the capsule covering the thyroid gland.

parietal (pah-ri'e-tal): relating to the walls of a space or cavity.

parturition (par-tu-rish'un): the process of giving birth to a child.

pelvis (pel'vis): any basinlike structure; an oblong trough; the lower portion of the trunk of the body bounded by the sacrum and coccyx at the back, the 2 hip bones at the sides and front, and the tissues of the pelvic floor at the outlet.

penis (pe'nis): the male organ of copulation.

pericardium (per-i-kar'de-um): the serous membrane that lines the sac enclosing the heart, plus the reflection that attaches itself to the heart itself.

perichondrium (per-i-kon'dre-um): a membrane that covers the surface of cartilage.

perineum (per-i-ne'um): the pelvic floor; the space between the anus and the scrotum in the male; the parts between the vagina and the anus in the female.

periosteum (per-e-os'te-um): the special fibrous connective tissue membrane covering the bones of the body; the surface tissue which plays an important part in the repair of bone fractures and other injuries.

peripheral (pe-rif'er-al): situated away from a center or central structure.

peristalsis (per-e-stal'sis): a rhythmic wavelike motion of the muscle tissue of the alimentary canal which moves the food through the digestive tube.

peritoneum (per-i-to-ne'um): the large serous membrane that lines the abdominal cavity and is reflected over the organs within.

phagocyte (fag'o-site): any cell that engulfs other cells, including bacteria, or any small foreign particles. Among phagocytes there are white blood cells plus certain cells in the spleen, liver, and lymph nodes.

pharyngeal (fah-rin'je-al): relating to the pharynx.

pharynx (far'inks): the saclike tube extending from the nose and mouth above to the larynx and esophagus below; the "throat."

phenylalanine (fen-il-al'ah-nin): a naturally occurring amino acid, found in milk and other foods, and required for normal growth and nitrogen balance.

physiology (fiz-e-ol'o-je): the science which deals with the activities or functions of the body and its parts.

pineal (pin'e-al): pertaining to the flattened cone-shaped body or glandlike organ near the midbrain.

pituitary (pi-tu'i-tar-e): the "master gland," which produces a number of hormones that regulate many body processes.

placenta (plah-sen'tah): a flat, circular organ which provides for the nutrition, respiration and excretion of the developing fetus.

progesterone (pro-jes'ter-one): a hormone produced in the female sex glands which assists in the normal development of pregnancy.

proprioceptive (pro-pre-o-sep'tiv): receiving stimulations within the body tissues, especially in muscles, tendons and in the inner ear.

prostaglandin (pros-tah-glan'din): 1 of a number of fatty acids which stimulate smooth muscle contractility, cause a lowering of blood pressure and affect the action of various hormones.

protein (pro'te-in): any of a group of complex organic compounds consisting of carbon, hydrogen, oxygen and nitrogen (some contain sulfur and phosphorus); the principal constituent of cell protoplasm.

prothrombin (pro-throm'bin): a substance in the blood plasma that is converted into thrombin during the second stage of blood clotting (coagulation).

protoplasm (pro'to-plazm): the living building material of all organisms, plants and animals; the only matter in which life is manifested.

proximal (prok'si-mal): near the point of origin; referring to the nearest part. The opposite is distal.

ptyalin (ti'ah-lin): an enzyme of the saliva which starts the digestive process by changing some starches into sugars.

pulmonic (pul-mon'ik): referring to or relating to the lungs, the pulmonary artery or the pulmonary valve.

recessive (re-ces'siv): the one of a pair of characters that will remain latent in the presence of the dominant gene, and appears only if both parents contribute this same gene.

renal (re'nal): pertaining to the kidney.

replication (re-pli-ka'shun): the process of producing a duplicate; a copying or duplication; said of DNA.

reticuloendothelial (re-tik-u-lo-en-do-the'le-al): pertaining to a combination of certain cells of the spleen, lymph nodes, bone marrow, and liver, that play roles in defense against infection, in metabolism, and in blood cell formation; pertaining to cells of both endothelium and reticulum.

reticulum (re-tik'u-lum): a network of connective tissue cells and fibers; a web-like mesh of protoplasmic cell extensions.

retina (ret'i-nah): the innermost coat of the eyeball; the nerve coat of the eye, made of nerve cells and fibers.

retroperitoneal (re-tro-per-i-to-ne'al): located behind the peritoneum as are the kidneys, pancreas and abdominal aorta.

ribonucleic acid (ri-bo-nu-kle'ik): the substance which carries DNA, the blueprint for all cells, to all parts of the body. Also called **RNA**.

roentgen (rent'gen): the international unit of radiation; a standard quantity of x or gamma radiation.

roentgenogram (rent-gen'o-gram): a film produced by means of x-rays, also called roentgen rays; named after the discoverer of the x-ray.

rugae (ru'ge): folds as of mucous membrane found in the lining of the stomach and elsewhere in the body.

sclera (skle'rah): the tough opaque white coat that forms the outer protective layer of the eyeball. It is continuous with the transparent colorless cornea at the front.

scrotum (skro'tum): a sac suspended between the thighs of the male which contains the testes.

sebaceous (se-ba'shus): secreting or pertaining to oil or an oily substance called **sebum** (se'bum).

secretin (se-kre'tin): a hormone produced by the stomach and small intestine which stimulates other digestive organs to produce their digestive juices.

secretion (se-kre'shun): the process of producing a new substance from materials in the blood; the new substance produced by glandular activity using materials in the blood.

semen (se'men): the thick whitish secretion from the male reproductive organs; a combination of male germ cells (spermatozoa) and secretions from the several glands of the reproductive system.

semipermeable (sem-e-per'me-ah-bl): permitting the passage of some particles (molecules) and not others; referring to membranes that allow the passage of a pure solvent such as water but not of the substances dissolved in it.

serosa (se-ro'sah): a serous membrane; one that is found lining the body cavities, such as the pleural and peritoneal cavities.

specificity (spes-i-fis'i-te): the quality of having a certain action affecting only a particular substance, tissue or organism.

sphygmomanometer (sfig-mo-mah-nom′e-ter): an instrument for measuring arterial blood pressure.

subclavian (sub-kla′ve-an): located under the clavicle (collarbone).

subcutaneous (sub-ku-ta′ne-us): occurring or located beneath the skin.

sudoriferous (su-dor-if′er-us): producing or secreting sweat.

sulcus (sul′kus): a groove or depression between parts, especially a fissure between the convolutions of the brain; any furrow, as in the teeth, the bones or in the lung surfaces. The plural form is **sulci.**

suprarenal (su-prah-re′nal): located above the kidney; relating to the adrenal gland.

suture (su′tur): a type of joint, especially in the skull where bone surfaces are dovetailed and closely united; a stitch used in surgery to bring parts and edges together.

symphysis (sim′fi-sis): a line of union; a cartilaginous joint such as that between the bodies of the pubic bones.

synapse (sin′aps): the region where parts of 2 neurons are anatomically related so that impulses are transmitted from one neuron to another. It is also called the **synaptic junction.**

synovial (si-no′ve-al): relating to a thick fluid found in joints, bursae and tendon sheaths.

systemic (sis-tem′ik): affecting the whole body; generalized.

systole (sis′to-le): the period of heart muscle contraction, especially that of the ventricles. The adjective is **systolic** (sis-tol′ik).

tensor (ten′sor): any muscle that stretches or pulls on a part to make it tense.

testosterone (tes-tos′ter-one): the hormone produced by the male sex glands.

tetany (tet′ah-ne): muscle spasms due to a low concentration of blood calcium.

thalamus (thal′ah-mus): the part of the brain at each side of the third ventricle, which acts as the chief relay center for sensory impulses to the cerebral cortex. It includes 2 large masses of gray matter.

therapy (ther′ah-pe): the treatment of disease or of any disorder.

thoracic (tho-ras′ik): relating to the chest portion of the body.

thorax (tho′raks): the chest; the part of the body between the neck and the abdominal cavity from which it is separated by the diaphragm.

thrombocyte (throm′bo-site): a particle of protoplasm found in the circulating blood; a blood platelet, believed to play a part in the process of blood clotting.

thromboplastin (throm-bo-plas′tin): a substance released by the injured tissues which triggers the clotting mechanism.

thymus (thi′mus): an elongated mass of lymphatic tissue, usually consisting of 2 lobes, located in the upper chest cavity beneath the sternum. It is believed to play a part in the immunity responses of the body.

thyrotropic (thi-ro-trop′ik): pertaining to an influence on the thyroid gland, as is the case with certain pituitary hormones.

thyroxine (thi-rok′sin): the hormone produced by the thyroid gland. It increases the metabolic rate and is needed for normal growth.

toxemia (toks-e′me-ah): a general toxic condition in which poisonous bacterial products are absorbed into the blood stream.

trachea (tra′ke-ah): a membranous and cartilaginous tube, commonly called the windpipe, extending from the larynx to its 2 branching bronchi.

tracheobronchial (tra-ke-o-brong′ke-al): relating to the trachea and the bronchi.

triceps (tri′seps): having 3 heads or 3 points of origin, especially in referring to the attachments of muscles.

tricuspid (tri-cus′pid): a valve in the right side of the heart that closes when the ventricle begins pumping.

trypanosoma (tri-pan-o-so′mah): a genus of protozoa. *Trypanosoma gambiense* causes African sleeping sickness.

trypsin (trip′sin): an enzyme in the gastric juice which splits proteins into amino acids.

tympanic membrane (tim-pan′ik): the eardrum.

umbilical (um-bil′i-kal): relating to the umbilicus or navel; also referring to the section of the abdominal cavity around the umbilicus in the central part of the abdomen.

umbilicus (um-bil′i-kus): a small scar on the abdomen which marks the former attachment of the umbilical cord to the fetus; the navel.

unilateral (u-ne-lat′er-al): pertaining to one side only.

urea (u-re′ah): a nitrogen waste product excreted in the urine; an end product of protein metabolism.

ureters (u-re′ters): in the urinary system, the 2 ureters conduct the secretion from the kidneys to the urinary bladder.

urethra (u-re′thrah): the excretory tube for the bladder.

uterus (u′ter-us): a muscular pear-shaped organ in the female pelvis within which the fetus grows until birth.

uvula (u′vu-lah): a soft, fleshy v-shaped mass which hangs from the soft palate.

vaccine (vak′sene): a substance used to cause antibody formation. Usually a suspension of attenuated or killed pathogens given by inoculation in order to prevent a specific disease.

vagina (vah-ji′nah): the lower part of the birth canal which opens to the outside of the body; the female organ of copulation.

varices (var′i-sez): enlarged veins; varicose veins. The singular form is **varix** (var′iks).

varicose (var′e-kose): pertaining to an unnatural swelling, as in the case of a varix or varicose vein.

vascular (vas′ku-lar): relating to or containing many vessels.

vector (vek′tor): a carrier of disease-producing organisms from one person to another (especially insects, such as mosquitoes).

venule (ven′ul): a very small vein that collects blood from capillaries.

vernix caseosa (ver′niks ka-se-o′sah): an oily substance resembling cream cheese which covers the skin of the fetus.

vertebra (ver′te-brah): any 1 of the 26 bones of the spinal column. The plural form is **vertebrae** (ver′te-bre), the adjective is **vertebral** (ver′te-bral).

villi (vil′li): tiny, fingerlike projections in the mucosa which lines the small intestine. The singular form is **villus**.

viscera (vis′er-ah): the organs in the 3 large body cavities, such as the stomach and the liver in the abdominal cavity. The singular is **viscus** (vis′kus).

zygote (zi′gote): the fertilized ovum, the cell formed by the union of the spermatozoon and the egg cell.

Index

J